A soaking pool
at Umpqua Hot
Springs
WAYNE ESTES

TOURING
HOT SPRINGS
WASHINGTON
AND OREGON

The States' Best Resorts and Rustic Soaks

THIRD EDITION

Jeff Birkby

FALCON GUIDES

GUILFORD, CONNECTICUT

FALCONGUIDES®

An imprint of The Rowman & Littlefield Publishing Group, Inc.
4501 Forbes Blvd., Ste. 200
Lanham, MD 20706
www.rowman.com

Falcon and FalconGuides are registered trademarks and Make Adventure Your Story is a trademark of The Rowman & Littlefield Publishing Group, Inc.

Distributed by NATIONAL BOOK NETWORK

British Library Cataloguing in Publication Information available

Library of Congress Cataloging-in-Publication Data

Names: Birkby, Jeff, author.
Title: Touring hot springs Washington and Oregon : the states' best resorts
   and rustic soaks / Jeff Birkby.
Description: Third edition. | Lanham : FalconGuides, [2020] | Revised
   ˉedition of: Touring hot springs Washington and Oregon: a guide to the
   states› best hot springs. | Includes bibliographical references and
   index. | Summary: "Whether you›re searching for a family hot springs
   resort or an isolated natural soaking pool, Touring Hot Springs Washington and
   Oregon will guide you to a truly memorable geothermal experience."— Provided by
   publisher.
Identifiers: LCCN 2020036246 (print) | LCCN 2020036247 (ebook) | ISBN
   9781493046645 (paperback) | ISBN 9781493046652 (epub)
Subjects: LCSH: Hot springs—Washington (State) —Guidebooks. | Hot
   springs—Oregon—Guidebooks. | Washington (State) —Guidebooks. |
   Oregon—Guidebooks.
Classification: LCC GB1198.3.W2 B57 2020 (print) | LCC GB1198.3.W2
   (ebook) | DDC 551.2/309795—dc23
LC record available at https://lccn.loc.gov/2020036246
LC ebook record available at https://lccn.loc.gov/2020036247

♾️™ The paper used in this publication meets the minimum requirements of American National Standard for Information Sciences—Permanence of Paper for Printed Library Materials, ANSI/NISO Z39.48-1992.

It would not be transcending the truth to state that all diseases are cured or largely benefited by the wonderful energy of these waters . . . health waits for those who may come and partake of this—life's elixir.

—Michael Earles, Sol Duc Hot Springs Resort, 1912

CONTENTS

The Hot Springs

WASHINGTON HOT SPRINGS

OREGON HOT SPRINGS

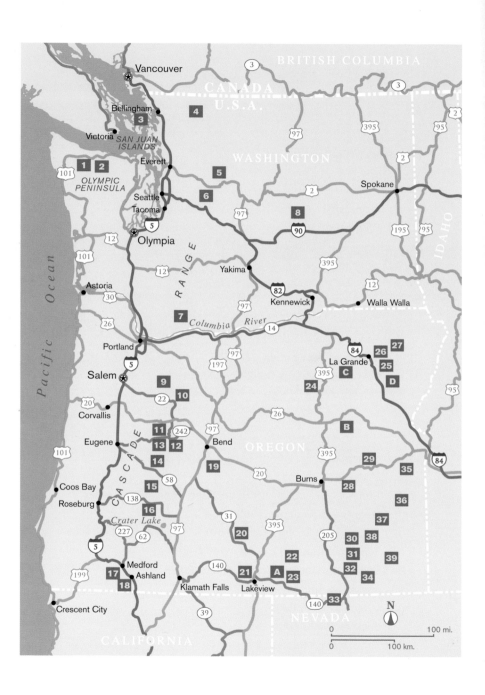

ACKNOWLEDGMENTS

Many of the quirky histories of Washington and Oregon hot springs would never have found their way into this book without the help of dozens of librarians, Forest Service and Bureau of Land Management personnel, museum archivists, and hot springs resort owners. My thanks to all the public servants and private hot springs owners who searched their memories and file cabinets to unearth the quotes and facts that fill these pages.

Thanks also to my many friends and family members who turned what might have been lonely solo trips into joyful adventures to Sol Duc, Breitenbush, Soap Lake, and many other soaking spots during my research for all three editions of this guide.

This third edition of the guide to Washington and Oregon hot springs is again dedicated to my parents, Robert and Evelyn Birkby, who have joined me on many a hot springs trip in the Northwest over the past four decades.

The author's parents and their grandson enjoy a rainy day soak at Sol Duc Hot Springs.

INTRODUCTION

Ask a dozen people on the streets of Portland or Seattle if they've ever been to one of Washington's or Oregon's hot springs, and you'll be lucky if you find a single person who can even name one of the region's thermal wonders, let alone one who has soaked there. Most residents of the Pacific Northwest know of the volcanoes that form the Cascade Range—Mount Rainier, Mount Baker, Mount Hood, Mount St. Helens, and several others. Few realize, however, that the geologic forces that created the region's volcanoes also left behind dozens of actively bubbling reminders of the tremendous thermal energy that underlies the two states.

The hot springs of Washington and Oregon are as varied as the region's topography. Although many hot springs are sprinkled throughout the Cascade Range, quite a few thermal soaks are also found in the rolling Blue Mountains of northeast Oregon, the dry southeastern Oregon deserts near the Nevada border, and the rain-drenched valleys of Washington's Olympic Peninsula.

Native American tribes in Washington and Oregon share legends of visiting the hot springs for curing their ills, and most hot springs were viewed as neutral territories where warring tribes could meet in peace to share the hot waters. Early pioneers reported seeing Native Americans building sauna huts on the shores of Soap Lake in central Washington and pouring mineral water from the lake onto heated rocks to create steam in the huts. Native Americans living on what is now the Warm Springs Reservation in north-central Oregon have bathed in local thermal waters for generations.

The first Europeans to see hot springs in Oregon and Washington were probably members of the Astor Expedition, who passed by the bubbling springs at Hot Lake in northeastern Oregon in 1812. Writer Washington Irving later retold these explorers' discovery of the region's first hot springs:

Emerging from the chain of Blue Mountains, they descended upon a vast plain, almost a dead level, sixty miles in circumference, of excellent soil, with fine streams meandering through it in every direction, their courses marked out in the wide landscape by serpentine lines of cotton-wood trees, and willows, which fringed their banks, and afforded sustenance to great numbers of beavers and otters.

In traversing this plain, they passed, close to the skirts of the hills, a great pool of water, three hundred yards in circumference, fed by a sulphur spring, about ten feet in diameter, boiling up in one corner. The vapor from this pool was extremely noisome, and tainted the air for a considerable distance. The place was much frequented by elk, which were found in considerable numbers in the adjacent mountains, and their horns, shed in the spring-time, were strewed in every direction around the pond.

By the 1840s thousands of settlers traveling west on the Oregon Trail had stopped at Hot Lake to rest and refresh themselves in the thermal water.

Most of the hot springs in Oregon and Washington had been discovered by the 1880s, and rustic bathhouses had been constructed at several. The peak of the region's hot springs popularity occurred during the first three decades of the twentieth century, when sumptuous resorts and sanitariums were constructed at hot springs throughout the region. Destination resorts built during this period included majestic Sol Duc Hot Springs Hotel on the Olympic Peninsula, the McCredie Springs Hotel in Oregon's Cascade Range, several hot springs resorts near Ashland in southwestern Oregon, and the resort hotels on the shores of central Washington's Soap Lake. Hot springs water was popular not only for bathing but also for drinking. Many of the early health resorts encouraged consumption of hot water. The town of Ashland, Oregon, still has water fountains downtown that supply mineral water to the public.

This consumption of hot water for health purposes was popularized in the only national park in the nation built around thermal water: Hot Springs National Park in Arkansas. A National Park Service brochure described the popular use of hot water for drinking:

Water from the hot springs is Hot Springs National Park's primary resource. Congress first protected the hot springs in 1832, and it intended for the water to be used.

Drinking the hot springs water is perfectly normal, even encouraged. Go ahead. "Quaff the elixir," as they used to say in the heyday of the spa (1880–1950 was the Golden Age of Bathing). Thousands of visitors highly endorse the good quality of the hot springs water and fill bottles to take home.

In the 1800s, springs were considered a practically guaranteed source of safe-to-drink water. Springs were categorized according to supposed mineral contents: sulphur springs, magnesia springs, chalybeate (or iron) springs, etc., and each type was considered medicinal for specific ailments. Spring waters acquire minerals by dissolving them out of rocks below the earth's surface.

EARLY WATER USE

The austerity imposed by World War II, the trend toward modern medical treatments, and several devastating fires led to the decline of Oregon and Washington's most famous resorts. Several of the elegant resort hotels had disappeared by the 1950s, with some, such as McCredie and Olympic Hot Springs, reverting to natural hot springs pools with no visible signs of their glorious resort past. A few of the original resorts from the early 1900s are still in operation, including the Ritter Springs Hotel, near the John Day River in northeastern Oregon, and the corrugated-steel bathhouse at Summer Lake Hot Springs, north of Lakeview, Oregon.

Since the mid-1980s there has been a resurgence of interest in the hot springs of Washington and Oregon, as city dwellers have increasingly sought escape from the pressures of urban life. New and revived resorts include one in the Columbia River

Gorge: Carson Hot Springs Golf & Spa Resort. The hotel at Hot Lake Springs near La Grande, Oregon, once abandoned and vandalized, has also been restored to its former elegance.

Oregon and Washington hot springs offer a variety of options for a quick getaway, including isolated soaking pools miles from civilization, family resorts crowded with weekend visitors from Seattle and Portland, holistic health centers deep in old-growth forests, and romantic inns with natural mineral-water hot tubs in cozy private rooms.

HOW TO USE THIS GUIDE

BOOK ORGANIZATION

Washington and Oregon offer an amazing variety of hot springs, as well as a number of ways to appreciate these thermal wonders. This book was written as a touring guide, focusing on both commercially developed and natural undeveloped hot springs in Oregon and Washington that are easily accessible and usually require no more than a short walk from your car to reach. Fortunately this requirement excludes only a handful of remote wilderness hot springs that entail more intensive hikes or raft trips to access.

A few of the hot springs described in this book are to be visited simply to savor their historical past, as they are currently closed and off-limits for soaking. Others are quiet, primitive pools where you can soak in solitude in a forested mountain valley or on an endless sagebrush prairie. Still other hot springs are found in quirky old resorts that look much the same as they did when they were built a century ago.

In some cases this guide does include descriptions of closed hot springs that have a compelling history and public access to vantage points. Medical Springs near La Grande, Oregon, and Crump Geyser near Lakeview, Oregon, are a couple of the off-limits but historic geothermal locations discussed in this guidebook. If you visit these areas, please respect private property boundaries and observe the hot springs from the public roads. If you want to enter private land to observe or soak in a hot spring, always ask permission first. (To determine land ownership, either ask nearby homeowners or check records at the county courthouse.)

Ownership situations constantly change with hot springs. Well-known soaking resorts, such as Blue Mountain Hot Springs and Lehman Hot Springs in eastern Oregon, have closed to public soaking. Meanwhile, some new and refurbished hot springs resorts have flourished since the two previous editions of this guide, including the Carson Hot Springs Golf & Spa Resort and Grande Hot Springs RV Resort.

This guidebook is divided into eight geographical sections. The Washington regions include the Olympic Peninsula, San Juan Islands, Washington Cascade Mountain Range, and central Washington. The Oregon hot springs are grouped into the Oregon Cascade Mountain Range, central Oregon, northeast Oregon, and southeast Oregon.

Each geographical section of the book presents detailed information on the hot springs in that area, with the chapters arranged to facilitate travel from one hot spring to the next for persons who want to plan tours to several hot springs in a single trip.

DIRECTIONS AND MAPS

The majority of hot springs in Washington and Oregon are near paved roads or highways and require little navigational skill to locate. Each chapter of this book contains detailed directions on finding the hot springs, accompanied by a clear map. Although most of the springs should be easy to find with the maps and directions in this book, you may want to obtain a topographical map before visiting some of the more isolated springs (topo maps can be obtained directly from the US Geological Survey, from map supply stores, or at many outdoor recreational stores). The DeLorme map books are also excellent (the DeLorme map page coordinates are included for each hot spring in this book). The following route abbreviations are used in the text: I (interstate highways); WA or OR (Washington or Oregon state highways); US (US highways); FR or CR (forest or county roads). GPS coordinates are also given for each hot spring, so you can easily enter the coordinates in your smartphone or GPS locator to zero in on your destination.

HAVE A SAFE HOT SPRINGS ADVENTURE

Although visiting the commercial hot springs resorts in Oregon and Washington requires no more preparation than you would take before traveling to any developed vacation spot, bathing in the more rustic or primitive pools requires a few precautions. Soaking in a natural hot spring is one of life's true pleasures, but be sure to plan for a safe experience.

Here's a quick checklist of things to remember before you head out on your next hot springs adventure:

- ❐ Always test the water temperature before you get into a thermal pool. Most people find soaks between 95 and 110 degrees F to be the most comfortable. At developed resorts in Washington and Oregon, you can usually assume that the water in the hot tubs and swimming pools is within this comfort zone. Some of the natural hot springs, however, approach 200 degrees F, so never take it for granted that a thermal pool will be a comfortable soak. Even hot springs in the same area may differ drastically in temperature. Some hot springs pools in the Alvord Desert of southeastern Oregon are cool enough for a pleasant soak, while other thermal pools mere yards away could easily poach an unwary bather. Even hot springs that you've visited before can have a significantly hotter or cooler temperature on subsequent visits. Some hot springs are much cooler early in spring and summer, when melting snows mix with thermal water. Play it safe before you soak—always test the water with your fingertip, then cautiously ease into the pool. The bottoms of some thermal pools are much hotter than the surface (and vice-versa), so be cautious even after you've settled in for your soak.

- ❐ Keep your head above water in natural hot-water pools. Some of the more popular soaking pools in Washington and Oregon have very sluggish flows, such as Baker and Olympic Hot Springs. When these pools are heavily used by soakers, the water flow may be insufficient to keep the bacterial count below water-quality standards. Drinking or inhaling this water can expose you to a variety of nasty bacteria. Try to avoid inhaling water droplets or spray, especially in primitive thermal pools.

- Don't soak by yourself. Soaking with friends is not only more enjoyable, it's safer.

- Drink plenty of (nonalcoholic) fluids. Hot-water soaks can increase your body temperature and put abnormal stress on your heart. Drink plenty of water during your hot springs bath, and avoid alcoholic beverages, especially if you've been soaking for a long time. Lengthy thermal soaks and a high alcohol level in your bloodstream make a dangerous combination—stick to nonalcoholic beverages and save the wine or beer until you get home from your soak.

- Don't soak in a hot pool for long periods of time if you're pregnant.

- Remove your jewelry before you get into thermal pools. The sulfur found in some hot springs may quickly tarnish your favorite ring or bracelet.

- Keep an eye on your kids. If you're soaking with children, make sure they are close by at all times, especially when soaking in thermal areas that have very hot pools or fragile crusts that a child could fall through.

- Watch out for poison ivy. The trails to several hot springs in Oregon and Washington are infamous for poison ivy infestations, especially those hot springs in the western parts of the states that are below 1,000 feet in elevation. If you happen to touch these shiny three-leafed plants, wash your skin surface immediately. It's best to wear a long-sleeved shirt and pants when walking in areas that might be infested. Hot spring soaking has been a popular treatment for poison ivy in the past—Carson Hot Springs Golf & Spa Resort in the Columbia River Gorge used to offer a regimen of thermal soaks to help relieve inflammation if you had a nasty encounter with poison ivy.

- Lock your valuables in your car trunk or, better yet, leave them at home. Unfortunately several hot springs in Oregon and Washington where your parked car is some distance from the hot springs have developed a reputation for vandalism. The Forest Service even has a name for the act of bashing in an automobile window and stealing your valuables—"car clouting." The parking areas at Bagby Hot Springs, Olympic Hot Springs, and Baker Hot Springs have been known in the past for this kind of vandalism. If you must take valuables with you on your trip, either lock them out of sight in your trunk or take them with you in a daypack to the hot springs and then keep a close eye on them while you're soaking.

CHECKLIST FOR A GREAT HOT SPRINGS VISIT

Preparing to visit the more luxurious hot springs resorts in Oregon and Washington requires packing little more than a swimsuit and some good books to read in the lounge chairs. But if you're planning on soaking in one of the more rustic or primitive thermal pools in the region, the following items may come in handy:

- ❏ US Forest Service or BLM maps, or a GPS locator/smartphone. A map is very useful for finding some of the more remote hot springs.

- ❏ Swimsuit. Although swimsuits are optional at some rustic hot springs, take one along just in case.

- ❏ A towel or two (one for drying yourself and one for standing on when changing your clothes).

- ❏ Daypack or plastic garbage sacks. Use to keep your clothes and towel dry while you're soaking. You can also use the garbage sack to pack out any trash you find when you leave a backcountry hot spring.

- ❏ Rubber thongs or old sneakers. Wear these to protect your feet from sharp rocks in pools.

- ❏ Plenty of drinking water. It's easy to get dehydrated if you soak for a long time.

- ❏ Snack food.

- ❏ Sunglasses.

- ❏ A full tank of gas. Be sure to fill up your gas tank, especially when visiting the remote hot springs in eastern Oregon. It can be a hundred or more miles between gas stations.

- ❏ A hat that provides good shade protection.

- ❏ Sunscreen. Use sunscreen on your nose, ears, and other areas not submerged. Refrain from using lotion on parts of your body that will be underwater, so you don't pollute the soaking pools.

- ❏ Skin moisturizer. Use after soaking.

- ❏ Flashlight. A flashlight is especially important if you're visiting some of the hot springs deep in the old-growth forests of the Oregon and Washington Cascades. Darkness falls quickly in these areas, and it's difficult and dangerous to try to find your way back to your car through the pitch-black forest.

- ❏ Business cards or pen and paper. You'll run into some wonderful people when you tour hot springs, and it's nice to share names and addresses with folks you meet.

- ❏ Northwest Forest Pass. Several of the hot springs in Oregon and Washington now require that you have a USDA Forest Service trail pass displayed in your car at the trailhead or carried with you. Other hot springs on public land may have a small day-use fee that can be paid in cash (no checks or credit cards). The hot springs information in each chapter of this book tells you what kind of pass you'll need and where to purchase it.

RESPONSIBLE BEHAVIOR

The vast majority of people who visit Oregon and Washington hot springs know that these thermal areas are rare natural wonders, and they treat both the hot springs and other visitors with courtesy and respect. Unfortunately, a few individuals haven't been as respectful of some of the more primitive pools, and their inconsiderate actions have resulted in restrictions on some popular hot springs areas. For example, Deer Creek Hot Springs, Terwilliger Hot Springs, McCredie Hot Springs, Snively Hot Springs, and several other public hot springs used to be open at night. But after years of alcohol-related incidents, Forest Service officials closed these springs to all after-dark bathing.

You can do your part to protect the fragile nature of hot springs areas as well as enhance the quality of the soaking experience for both yourself and other visitors by keeping in mind a few rules of etiquette during your visit:

- ❐ Don't bring glass containers. Nothing ruins a hot springs soak faster than cutting your feet on a shard of broken glass.

- ❐ Pack out all your trash. And pack out as much trash as you can that's been left by others.

- ❐ Respect private property rights. Don't enter a hot spring on private land without being sure the owner allows public access.

- ❐ Keep noise to a minimum. Soaking in hot springs is an almost mystical experience to some bathers. Loud parties or blaring radios ruin the mood for everyone.

- ❐ Watch where you walk. Some of the hot springs in Oregon and Washington have deposited beautiful ledges of fragile minerals, which may have taken decades to build up. Walking on these delicate areas can destroy these deposits. It's also possible to break through this fragile crust, which hides scalding water just below the surface. Be especially careful when walking near the beautiful hot springs in the Alvord Desert of southeastern Oregon.

- ❐ Follow local conventions on nudity. Several hot springs in this book are clothing optional, but use your own judgment before you strip. If you come to a primitive hot pool that's already occupied by clothed bathers, then either ask if it's okay to soak nude or follow the majority lead and wear your swimsuit.

WHAT'S MISSING, WHAT'S SPECIAL

Hot Springs Not Included in This Guidebook

There are more than one hundred hot springs and hot-water wells in Oregon and Washington. Unfortunately many of these thermal sources are on private land that is closed to the public, are too hot for comfortable soaks, or don't produce enough water to make a decent soaking pool. Most of these hot springs aren't included in this guidebook.

This guidebook focuses on the most accessible hot springs in Oregon and Washington, those that don't involve serious hiking or off-road driving. It includes both commercial resorts and secluded natural hot springs, but all these thermal areas are close to roads or are, at most, an easy walk from your vehicle.

AUTHOR'S FAVORITE HOT SPRINGS

Most Remote: Willow Creek Hot Springs

It may be a contradiction to feature a "most remote" hot spring in a guidebook that emphasizes easily accessible thermal soaks, but Willow Creek Hot Springs in southeastern Oregon is both accessible and remote. Reaching this soak in the high desert of southeast Oregon requires a long day's drive from almost anywhere in the state, and the final few miles of dirt road can be impassable in wet weather. Visitors who endure the bumpy journey will be rewarded with twin soaking pools surrounded by an endless sagebrush prairie.

Most Historic: Hot Lake Springs

A majority of hot springs in Washington and Oregon have interesting histories, but Hot Lake Springs in northeastern Oregon certainly has the oldest pedigree. First visited by European explorers with the Astor Expedition in 1812, Hot Lake Springs later became a popular stop for weary travelers on the Oregon Trail in the 1840s. Some historians consider the hotel built at Hot Lake Springs in 1864 to have been the first commercial building in the United States to use geothermal energy for space heating. The massive brick building, which sat abandoned for decades, was renovated and reopened in 2006, with much of the character and elegance from its past restored.

Best for Families: Sol Duc Hot Springs Resort

Tucked into the Olympic Peninsula in western Washington, Sol Duc Hot Springs Resort is an easily accessible hot springs vacation for the whole family. The hot and cold pools delight all ages of soakers, and the cabins with kitchen facilities help keep costs down when feeding a hungry family. Nearby RV and tent camping options also appeal to thrifty families. Nearby attractions include the Hoh Rain Forest Visitor Center and Sol Duc Falls.

Best Nude Soak: Doe Bay Resort and Retreat

Several hot springs in Oregon and Washington are clothing optional, but Doe Bay Resort and Retreat on Orcas Island gets a nod for the best bath in your birthday suit. The mellow atmosphere and well-maintained soaking facilities lend a sense of security to nude bathers that is sometimes lacking at other hot springs. Doe Bay features three soaking tubs, a sauna big enough to hold two dozen of your swimsuit-free friends, and a secluded nude beach where a curious sea otter in Puget Sound may be the only neighbor likely to catch a glimpse of your bare buns.

Most Romantic: Lithia Springs Resort

Combine private, two-person whirlpools in the cozy cottages at Lithia Springs Resort with nearby gourmet restaurants, and add the world-famous Oregon Shakespeare Festival, and you have a perfect recipe for romance. Honeymooners and retired couples alike credit the inn's silky-soft mineral water for turbocharging their relationships. One couple wrote in the inn's guest book that their "carnal knowledge" had been reawakened by the thermal springs. The many cultural attractions in nearby Ashland add perfect venues for building a weekend of memories with your significant other.

Most Eclectic: Soap Lake

The dry prairies of central Washington seem a strange place to find a once-booming spa town, but Soap Lake is filled with such eccentricities. The mineral-rich waters of the town's namesake have been praised for their abilities to cure everything from snakebites to baldness (not to mention the legend of the mineral water resurrecting a dead cowboy). A sizable contingent of immigrants from eastern Europe take up residence near the lake during the summer, both to soak in and to drink the celebrated water. The citizens of Soap Lake are constantly dreaming up ways to boost their small town, promoting everything from a midsummer health fair and canoe race to community mud baths on the lakeshore.

Map Legend

5	Interstate Highway
97	US Highway
22	State Highway
46	County/Forest Road
	Local Road
	Unpaved Road
	Trail
	Railroad
	Power Line
	State Border
	International Border
	River/Creek
	Intermittent Stream
	Body of Water
	Intermittent Lake
	Sand
	National Forest/National Park
	National Monument/Wilderness Area
	National Wildlife Refuge
	Indian Reservation
	State Park/City Park
	Bridge
	Campground
	Capital
	Dam
12	Featured Hot Spring
	Unfeatured Hot Spring
	Mountain/Peak (or elevation)
	Pass
	Picnic Area
	Point of Interest
	Ranger Station
	Town
	Tunnel
	Visitor Center
	Waterfall

WASHINGTON HOT SPRINGS

Overnight cabins at Sol Duc Hot Springs LINDSEY GUBLER

OLYMPIC PENINSULA

1. SOL DUC HOT SPRINGS RESORT

General description: A commercial resort nestled in a valley of the Northern Hemisphere's only temperate rain forest.

Location: Olympic Peninsula, 40 miles west of Port Angeles on the banks of the Sol Duc River in Olympic National Park.

Development: Sol Duc Hot Springs has been developed commercially over 100 years. Current facilities include a restaurant, cabins, massage therapists, poolside delicatessen, gift shop, 3 hot soaking pools, and a cold-water swimming pool.

Best time to visit: The resort is open from late Mar through mid-Oct. Because Sol Duc is in a temperate rain forest, expect cloudy and wet weather anytime. July and Aug are usually the driest (and busiest) months. Midweek is less crowded, as are the periods before Memorial Day and after Labor Day. Cabins are often booked weeks in advance, so make your reservations as early as possible.

Restrictions: Swimsuits are required in the swimming and soaking pools. Cabin fees include use of the pools; other guests must pay a day-use fee. There's a required 2-night minimum stay during holidays. The cabins and lodge are nonsmoking.

Access: Any vehicle can travel the paved highway and blacktop county road to the hot springs.

Water temperature: The 3 soaking pools are kept between 101 and 105 degrees F. The cold-water swimming pool averages 78 degrees F.

Nearby attractions: Sol Duc Hot Springs is a great location for accessing some of Olympic National Park's nicest views and hiking trails. Sol Duc Falls is about a mile past the resort and campground. Stop at the overlook to see this 60-foot waterfall dropping into a narrow canyon. One short (but strenuous) hiking trail starts at the resort and climbs about 2.5 miles to trout-filled Mink Lake. Another memorable hike that anyone can take is on the Ancient Groves Nature Trail, an easy 0.5-mile loop through a giant Douglas fir forest. The trail starts about 3 miles north of Sol Duc Hot Springs on the road leading back to US 101. For more insight into the ecology of the rain forest, head back to US 101 and drive west to the Hoh Rain Forest Visitor Center.

Services: The resort has an espresso bar, a gift shop, and a small delicatessen in the main lodge. The Springs Restaurant provides breakfast and dinner, and poolside lunches are served. Massage therapy is available. Sol Duc has 32 cabins available, some with kitchens. The cabins with kitchen facilities are closer to the lodge. Many guests choose to stay in the more remote cabins without kitchen facilities and eat their meals in the lodge restaurant (oatmeal with fresh berries for breakfast and steelhead trout for dinner are favorite choices). The resort also has 17 RV spots across the Sol Duc River from the cabins.

Camping: Tent camping is available less than a mile away at the Sol Duc Campground, maintained by the National Park Service. The campground has more than 80 tent sites and is available on a first-come, first-served basis, so arrive early in the day to grab a spot for your tent. During summer park rangers sometimes give an evening lecture at the campground on the natural history of the surrounding area. The Log Cabin Campground with 32 RV sites and 10 tent camping sites is located about a quarter mile from the resort—it's a great option if the Sol Duc Campground is full.

Maps: Washington State Highway Map; *DeLorme: Washington Atlas & Gazetteer,* page 43, A6.

GPS coordinates: N47.9700' / W123.8625'

Contact info: Sol Duc Hot Springs Resort, 12076 Sol Duc Hot Springs Rd., Port Angeles, WA 98363; (888) 896-3818; www.olympicnationalparks .com/accommodations/sol-duc-hot-springs-resort.aspx; info@visitsolduc .com.

Finding the springs: From Port Angeles drive 27 miles west on US 101. (The final 10 miles of this drive skirt the shores of Lake Crescent.) Turn left onto Sol Duc Valley Road (you'll see a large sign for Sol Duc Hot Springs at the junction with US 101). Drive 12 miles on this 2-lane blacktop road through beautiful groves of fir and cedar to the Sol Duc Hot Springs Resort.

THE HOT SPRINGS

All of Sol Duc Resort's soaking and swimming pools are clustered behind the main lodge. (Diners in the Springs Restaurant have a front-row view of the bathers.) Two circular hot pools, both about 20 feet in diameter, are the focus for most visitors. The pools, which are about 2 feet deep, are filled with natural hot water that measures 101 to 105 degrees F. A hot-water fountain gurgles from a platform in the center of the pool closest to the lodge. A smaller wading pool about 6 inches deep lies adjacent to the soaking pools. The hot pools are drained and cleaned every night after closing.

Farthest from the lodge is the cold-water swimming pool, which is 100 feet long by 30 feet wide and 2 to 10 feet deep. This chlorinated pool usually measures a brisk 74 to 78 degrees F. The chilly swimming pool water is especially popular with children, and adults can keep an eye on their clans from the comfort of the warm soaking pools.

A covered overhang near the dressing rooms and showers provides a dry place for towels. You'll probably need to use that dry area, as Sol Duc typically receives 150 inches of rain a year—that's more than 10 feet of annual rainfall. Some visitors to Sol Duc have been known to open an umbrella above them while soaking in the hot pools, keeping the cold drizzle off their heads and shoulders while they contently steam beneath.

History

Theodore Moritz was the first European to discover Sol Duc Hot Springs. During a hunting trip in the 1880s, Moritz came upon a Native American in the woods who had broken his leg. Moritz nursed the man back to health, and he returned the favor by telling Moritz of hot springs deep in the woods that were used by members of his tribe to treat their illnesses. Moritz followed the directions to the hot springs, where he saw

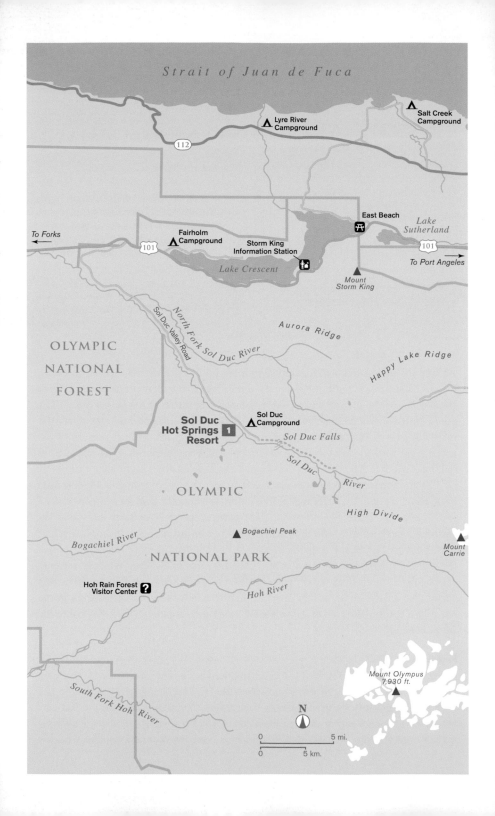

Strait of Juan de Fuca

Lyre River
Campground

Salt Creek
Campground

112

East Beach

Lake
Sutherland

To Forks

Fairholm
Campground

101

Storm King
Information Station

Lake Crescent

101

To Port Angeles

Mount
Storm King

Aurora Ridge

OLYMPIC

Happy Lake Ridge

NATIONAL

North Fork Sol Duc River

Sol Duc Valley Road

FOREST

Sol Duc
Hot Springs
Resort 1

Sol Duc
Campground

Sol Duc Falls

Sol Duc

River

OLYMPIC

High Divide

Bogachiel River

Bogachiel Peak

Mount
Carrie

NATIONAL PARK

Hoh Rain Forest
Visitor Center ?

Hoh River

South Fork Hoh River

Mount Olympus
7,930 ft.

N

0   5 mi.

0   5 km.

Waterfall near Sol Duc Hot Springs I INDSEY GUBLER

many tribesmen bathing in the hot water. Moritz filed a claim on the land and built a crude trail from the hot springs to Lake Crescent, then settled in on his new property, building a cabin near the springs and carving a soaking tub from a cedar log. Word of the hot springs spread to other settlers, and Moritz hosted a steadily increasing number of visitors who came to soak in the healing waters.

One of the visitors to Moritz's homestead was Michael Earles, a wealthy lumber baron from Seattle. In 1903 Earles was diagnosed with a life-threatening disease, and physicians had little hope for his survival. One doctor suggested that Earles travel to the famous Carlsbad hot springs in Czechoslovakia for treatment. Earles's health was failing rapidly, and he felt that he was too weak to make the European journey. He had heard of the rustic hot springs resort on the Olympic Peninsula and decided to see if the springs would be helpful for his malady. After a few weeks of soaking in the hot springs at Sol Duc, Earles emerged "as good as new," according to one newspaper article. Earles was so impressed with the curative power of the hot springs that upon Theodore Moritz's death in 1910, he purchased the property.

Earles transformed the little homestead into what may have been the most lavish resort in the Pacific Northwest. The main attraction of Earles's new resort was the Sol Duc Hot Springs Hotel, which opened in 1912. The four-story hotel was 160 feet long, 80 feet wide, and surrounded on three sides by a veranda 20 feet deep and 400 feet long. The building featured 165 bedrooms, a laundry room, an ice plant, and its own power plant. A massive stone-and-brick fireplace, an electric organ, and a dining room that seated 150 people greeted guests entering the main lobby. Living fir trees served as the main supports for the hotel. A ballroom provided a forum for orchestras that Earles occasionally invited to the resort. According to a dedication pamphlet from 1912, guests at the hotel relaxed while "music for turkey trots, tangos, or waltzes floated out to the long veranda on summer evenings as merrymakers splashed in the nearby pools while the sedate sat in chairs, gazing at the background of stars, mountains, and river."

Connected to the hotel by a covered walk was a three-story sanatorium, staffed by a resident physician, several nurses, and massage therapists and attendants. The sanatorium attracted the sick and infirm, who would spend weeks under the care of the physician. Next to the sanatorium was a bathhouse for recreational soaking and swimming. A gymnasium and several cottages were also scattered on the well-manicured grounds. Golf links, tennis courts, croquet fields, a movie theater, bowling alleys, and billiard rooms were available for the pleasure of hotel guests. Each resort guest was given his or her own cup to drink the mineral water from a bubbling hot springs fountain.

Earles promoted the resort to the Seattle community as "The Carlsbad of America." In a 1912 brochure announcing the opening of the new hotel, Earles lauded the miraculous healing qualities of the thermal water:

> The rheumatic diathesis, in common with hepatic, gastric, renal and every form of blood and skin diseases, each and all, succumbed with remarkable

Hot soaking pool at Sol Duc Hot Springs LINDSEY GUBLER

Overnight cabins at Sol Duc Hot Springs LINDSEY GUBLER

rapidity to the benignant energies of the water . . . it would not be transcending the truth to state that all diseases are cured or largely benefited by the wonderful energy of these waters . . . health waits for those who may come and partake of this—life's elixir.

The elegance of the Sol Duc Hot Springs Hotel was short-lived. Four years to the month after its opening, the hotel burned to the ground when sparks from a defective flue caught the wooden shingles on fire. A strong west wind blew the flames to most of the other buildings on the property, destroying the sanatorium, bathhouse, and cottages. One guest at the hotel reported that the fire short-circuited the electric organ, which then played Beethoven's "Funeral March" until the fire finally burned through its electrical wires. Within three short hours the raging fire had destroyed the resort.

The tragic fire of 1916 ended Sol Duc's fame as one of the most elegant destination resorts for wealthy patrons from Seattle. Michael Earles died three years after the fire, and the property was sold soon thereafter. Over the next fifty years, the resort had several owners who rebuilt the pool and some of the cottages, but the grandeur of the hotel and sanatorium of 1912 was never again realized.

In 1966 the owner of the resort sold the 320 acres, which included the hot springs, to the National Park Service for $880,000. Since then Sol Duc Hot Springs has been operated by private concessionaires on long-term leases from Olympic National Park.

2. OLYMPIC HOT SPRINGS

General description: A half-dozen wilderness soaking pools sprinkled along a hillside above Boulder Creek in Olympic National Park.

Location: Olympic Peninsula, 21 miles southwest of Port Angeles in Olympic National Park.

Development: Olympic Hot Springs was at one time a popular commercial resort, but all remnants of development were removed by the National Park Service in the 1970s. Volunteer-built rock-and-log dams enclose the current natural soaking pools.

Best time to visit: Weekends and holidays are often very busy—it's not uncommon to see 30 or 40 cars in the trailhead parking area. Winter may be the best time to visit, when the crowds are few and the steaming water provides a dramatic contrast to surrounding snowbanks. The soaking pools are often a bit cool in spring-time, when cold water from snowmelt mixes with the thermal water.

Restrictions: Because the hot springs are located within Olympic National Park, you'll need to purchase a national park pass (purchase at the ranger station near the mouth of the Elwha Valley). Camping permits are also required if you plan to stay overnight at the camping area near the hot springs. Camping is allowed only in the designated campsite. The National Park Service doesn't condone nudity in the soaking pools, but it isn't banned. Actually the Park Service doesn't condone soaking of any kind in the hot springs, because the sluggish water can contain high levels of coliform bacteria. Visitors routinely ignore both warnings, and Park Service personnel tolerate both bathing and nudity. There's usually a mix of clothed and nude bathers in the soaking pools.

Access: Any vehicle can make the trip to the trailhead for Olympic Hot Springs, except for a few weeks in winter when snow blocks the last 4 miles to the trailhead. Cross-country skiers are the main winter visitors when snow blocks the lower road. Bicycles are allowed on the 2.4-mile section of trail to the Boulder Creek Bridge, so consider bringing your mountain bike along. Since 2014 there hasn't been automobile access to Olympic Hot Springs via Olympic Hot Springs Road, due to washouts of sections of the access road. Check with the National Park Service for current road conditions.

Water temperature: The temperature in the soaking pools varies from 85 to 105 degrees F.

Nearby attractions: Olympic National Park has a wide variety of outdoor recreation opportunities. Stop at the main Olympic National Park Visitor Center in Port Angeles to plan your trip. One of the most popular summer drives is the 17-mile road up to Hurricane Ridge, one of the highest spots in the park, with awesome views and some nice day hikes. The Hoh Rain Forest Visitor Center is also popular. Located on the west side of the park, this visitor center has wonderful displays telling the story of the only temperate rain forest in the Northern Hemisphere. Nearby are easy hikes through moss-covered cedars and firs.

Services: No services are available at the hot springs. Groceries, gas, and lodging are available in Port Angeles,

21 miles northeast. Accommodations are available in Port Angeles. Sol Duc Hot Springs Resort, 30 miles west of Olympic Hot Springs off US 101, has cabins available for rent from mid-Mar to mid-Oct.

Camping: Boulder Creek Campground is 0.25 mile north of the hot springs. A camping permit is required.

Maps: Olympic National Park map and Olympic Hot Springs & Camping Area leaflet (both available at Elwha Ranger Station); *DeLorme: Washington Atlas & Gazetteer,* page 43, A8.

GPS coordinates: N47.9767' / W123.6886'

Contact info: Olympic National Park Wilderness Information Center, 3002 Mt. Angeles Rd., Port Angeles, WA 98362; (360) 565-3100; www.nps.gov/olym/planyourvisit/boulder-creek-trail.htm.

Finding the springs: From Port Angeles drive west on US 101 for 10 miles to Elwha River Road (also called Olympic Hot Springs Road). Turn left and drive to the entrance gate for Olympic National Park. The gate isn't always staffed, but if it is, pay the national park entrance fee and pick up a brochure about the trail to Olympic Hot Springs. Park your car at the Madison Falls parking lot. In past years you could drive about 8 miles past the parking lot, which brought you within 2 miles or so of the hot springs. But flooding and washouts just past the Madison Falls parking lot have made the road impassable to vehicles. So it's a long bike ride or hike from your car to the hot springs—about 10 miles one way. This makes for a very long day if you're hiking, but the abandoned road can be biked, so consider this option too. From the entrance gate to Olympic National Park, you'll need to hike or bike 8 miles on a winding road that ends at the parking area for the Appleton Pass Trailhead (the last 4 miles of this road switchback up a steep hill from Mills Lake). From the Appleton Pass Trailhead, it's a 2.2-mile hike on an abandoned blacktop road to the Boulder Creek Campground. Cross the footbridge over Boulder Creek (stop and look upstream for a great view of a waterfall), turn left, and follow the trail that parallels Boulder Creek downstream. In less than 5 minutes, you'll start seeing the soaking pools along the side of the trail.

THE HOT SPRINGS

Locals used to call the area "Triple 21 Hot Springs"—2,100 feet in elevation, 21 miles from Port Angeles, and twenty-one hot springs seeping out of the hillside above Boulder Creek. Those twenty-one hot springs seeps still exist, but only seven of them are collected into soaking pools. You'll encounter these seven pools every 50 to 100 yards after crossing the footbridge and taking the path to the left. Sometimes the hot pools will be obvious (only a few yards off the trail), but a couple of them are hidden higher up the ridge. Look for footpaths heading up the hill and little hot-water creeks to give you hints as to where these hidden pools might be.

The first soaking pool appears just above the hiking trail about 100 yards from the footbridge. Rocks and logs enclose the spring, which is about 6 feet wide by 15 feet long and 1.5 to 2 feet deep. Hot springs bubble up in this pool, as well as trickle in from a rock fissure. A river birch shades most of the pool, which has a rock-and-silt bottom.

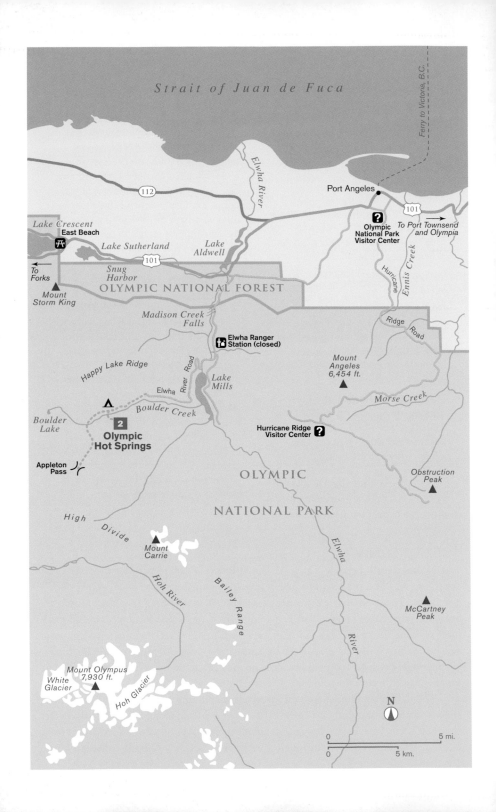

One of the picturesque soaking pools at Olympic Hot Springs

Don't be too hasty to jump in the first pool you see—the best soaking pools are those farthest down the trail. Past the first pool, on the left close to the creek, is a small soaking pool (the "meadow pool"). This pool is often overlooked because it's the only one not on the uphill side of the trail.

Uphill from the trail past the meadow pool are three other small soaking pools, each about 30 yards above (and hidden from) the trail. The sixth pool is actually a double family-size soaker right by the trail. These twin pools are about 20 feet across. The last pool (and to many folks the best) is past the family pools. Walk about 50 yards and then turn on the steep footpath headed up the ridge.

The trail climbs about 300 feet to a soaking pool in the middle of a hot water creek dammed with logs and rocks. This 6-by-15-foot pool holds a half dozen or more bathers and tends to be the hottest (and most popular) soaking pool in the series.

The hot springs water entering the seven pools tends to stay at a fairly constant temperature, but cold water from snowmelt and heavy rainstorms can dilute the pools. In rare cases during spring, this cold-water runoff can cool the pools to the point that they are uncomfortable for soaking, but most of the year this isn't a concern.

Rangers for Olympic National Park sometimes patrol the area, picking up trash and seeing that partying is kept under control. But since the road has washed out near the park entrance, the patrols by park personnel have become few and far between. It can still be a bit disconcerting the first time you see park service personnel in their uniforms and badges standing next to a pool full of naked bathers, but everyone seems to be pretty mellow about the situation.

History

Native Americans in the area have long had a legend about the creation of both Olympic Hot Springs and nearby Sol Duc Hot Springs:

> Two "dragon-like creatures" with a mutual hatred for one another engaged in a mighty and desperate battle. There was no victor as both were evenly matched. Admitting defeat, each of the creatures crawled into their separate caves where they still weep hot tears of mortification. The hot tears of these dragons are said to be the hot water bubbling to the surface in these two hot springs.

The first European to discover Olympic Hot Springs was Andrew Jacobsen, who stumbled across the steaming waters in 1892. Jacobsen was alarmed by the sight of the hot water, fearing it was a sign that the mountain was going to blow up. He beat a hasty retreat down the Elwha Valley away from the hot springs. Apparently no one else visited the site for the next fifteen years, until William Everett rediscovered the hot springs in 1907. Unlike Jacobsen, Everett had no fear of the mountain erupting and filed a homestead claim on the hot springs property.

Everett built a cabin and bathhouse near the springs and moved his family to the little homestead. In spite of the arduous 12-mile hike or horseback ride from the Port Angeles road to the hot springs, visitors soon started arriving at Everett's small bathhouse to soak in the hot water. In 1908 Everett decided to build some tent frames and rent overnight accommodations to the visitors. He also built a larger bathhouse containing six wooden tubs that he carved out of cedar logs. A lodge with a dining room was built in 1917, followed by a swimming pool 75 feet long and 25 feet wide. Cabins were added to the property in 1920, the same year that Everett's daughter married Harry Schoffel. Schoffel took over management of the resort in 1924 and spent six years building an even larger lodge on the property, complete with a dining room, a kitchen, and ten sleeping rooms. An additional ten sleeping rooms were added in 1932.

Visits to Olympic Hot Springs increased dramatically in 1930, when the Forest Service built 12 miles of new road from the highway to the little resort. Gone were the days when visitors had to walk or ride on horseback to get to the hot springs. Schoffel built an Olympic-size swimming pool and additional cabins near Boulder Creek behind the lodge to handle the increased visitor load. Resort business boomed for the next decade.

In January 1940 tragedy struck the resort when the lodge caught fire. The lodge was constructed almost entirely of cedar, and the fire quickly destroyed the building. Fortunately the fire didn't spread to the swimming pool or cabins, but it did mark the beginning of the end of Olympic Hot Springs' commercial success.

Shortly after the fire the federal government acquired the property and incorporated it into Olympic National Park. The resort was leased back to Schoffel, but new federal requirements were put in place. The National Park Service required Schoffel to use only chlorinated fresh water in the swimming pool, which drastically reduced the number of tourists who had previously visited the resort to soak in natural unchlorinated hot springs water.

The Schoffel family continued to manage the resort until 1966, when the National Park Service decided not to renew their lease. The resort was closed, and the abandoned buildings eventually collapsed under the weight of heavy winter snows. In 1972 the National Park Service decided to let the area revert to its natural state and removed all buildings from the area. At present the only signs that Olympic Hot Springs was once a bustling resort are a few cisterns and pieces of iron pipe near some of the springs that were used to gather the hot water for use in the old swimming pool.

SAN JUAN ISLANDS

3. DOE BAY RESORT AND RETREAT

General description: A rustic New Age retreat with clothing-optional soaking tubs overlooking a peaceful cove in Washington's San Juan Islands.

Location: Southeast end of Orcas Island, 19 miles from the Orcas ferry dock.

Development: The 45-acre property was originally developed as a ferry landing in the late 1800s. Current developments include a variety of sleeping accommodations, a general store, a restaurant, a massage house, a sauna, and soaking tubs.

Best time to visit: Doe Bay Resort and Retreat is open year-round, although the Doe Bay Café may be closed during winter months (call ahead for current status). According to a sign near the soaking area, the hot tubs and sauna are open "9ish a.m. to 10ish p.m." Weekends, holidays, and summer are busiest. Call well in advance of your trip for cabin reservations or to reserve one of the better camping spots next to Puget Sound. It's best to take a ferry early in the day from Anacortes to Orcas Island to ensure you arrive at the resort during daylight hours—the heavily forested resort is not well lit, and it can be difficult to find your assigned cabin or camping spot in the darkness.

Restrictions: The use of the mineral-water soaking tubs and the sauna is restricted to overnight guests and to visitors who pay a day-use fee. The hot tub and sauna area is clothing optional—most guests soak in the nude. A clothing-optional beach dubbed "little Miami" is also available. Nudity is not permitted outside the nude beach, hot tubs, or sauna area. No minors under the age of 18 are allowed in the soaking tub area after 6 p.m. No pets are permitted from June 15 to Sept 15. No dogs, alcohol, cigarettes, or glass containers are allowed in the hot tub area. No smoking is allowed in the cabins.

Access: Any vehicle can make the trip on the paved roads that lead from the ferry terminal across Orcas Island to Doe Bay Resort and Retreat.

Water temperature: Two hot tubs contain mineral water that's piped from a 300-foot-deep well, then heated to 105 to 110 degrees F. The adjacent cold-water pool is a bracing 52 degrees F. While the pools at Doe Bay aren't filled with water that is naturally heated, the gorgeous views and soothing warmth of the pools make it easy to forget this fact.

Nearby attractions: Horseshoe-shaped Orcas Island is the largest island in the San Juans, encompassing 54 square miles. Moran State Park, the largest state park in Washington, is situated 6 miles north of Doe Bay Resort and Retreat. A popular day trip is the hike or drive to the summit of Mount Constitution, which towers 2,409 feet above sea level and features an awesome view of the San Juan Islands and Olympic and Cascade Mountains. Elegant Rosario Resort, on Cascade Bay northwest of Moran State Park, features a 35,000-square-foot mansion built by the ex-mayor of Seattle, Robert Moran. Free concerts are often given at the mansion on a majestic 1,972-pipe organ. Eastsound, the largest community on Orcas Island, is home to a nice selection of restaurants.

Services: If you're seeking fancy accommodations, head down the road about 10 miles from Doe Bay to the elegant Rosario Resort. But if you want to revitalize your spirit in a rustic natural setting, then Doe Bay Resort and Retreat may just be your place. A small selection of groceries can be found in the store just off the lobby of the Doe Bay Café. A variety of massage therapies can be scheduled with the resident professional therapist. Guided sea-kayak tours of Doe Bay are offered (including whale-watching), and mountain-bike rentals are available.

Camping: Choose from a variety of sleeping options, which include primitive campsites with views of Otter Cove, octagonal yurts with skylights and futon beds that would be the envy of Mongolian tribesmen, and simply furnished cabins (some with kitchens). You can also join fellow travelers in a group hostel. RV spots are available too. Two community bathrooms and shower houses are provided for guests at campsites and in those cabins without toilets. Guests without cooking facilities can use a community kitchen. Guests often bring their own food to the resort and then write their names on their perishables before placing them in the community refrigerator with the provisions of other visitors.

Maps: The Washington State Highway Map is adequate, but you might want to pick up a more detailed map of Orcas Island in one of the free visitor guides available in the information racks on the Washington ferries.

DeLorme: Washington Atlas & Gazetteer, page 15, E6.

GPS coordinates: N48.6411' / W122.7810'

Contact info: Doe Bay Resort and Retreat, 107 Doe Bay Rd., Olga, WA 98279; (360) 376-2291; www.doebay .com; office@doebay.com. Ferry schedules to and from Orcas Island, (888) 808-7977 or (206) 464-6400; www.wsdot.wa.gov/ferries/.

Finding the springs: From I-5 north of Mount Vernon, take exit 230 and head west on WA 20 to Anacortes. Follow the signs in Anacortes to the San Juan Island ferry. The ferry ride from Anacortes to Orcas Island usually takes about an hour, during which you'll have a wonderful view of several of the islands of the San Juans. Once the ferry lands at Orcas Island, take the Horseshoe Highway for about 8 miles to the town of Eastsound. (Keep alert on this drive: Orcas Island is popular with bicyclists and deer, both of which can suddenly appear on the narrow winding roads.) Drive through Eastsound, proceeding east for about a mile until you come to Olga Road. Turn right onto Olga Road and drive 10 miles south to the entrance to Moran State Park. Stay on Olga Road and drive the 2 miles through the state park (you'll pass through a narrow arch at the park boundary). Continue beyond the state park on Olga Road for another 2.5 miles and turn left at the Olga Café onto Pt. Lawrence Road. Drive 3.5 miles on Pt. Lawrence Road until you see the Doe Bay Resort sign on the right. Turn into the resort and drive about 50 yards to a parking area near the Doe Bay Café. Park your car and go into the lobby of the cafe to register.

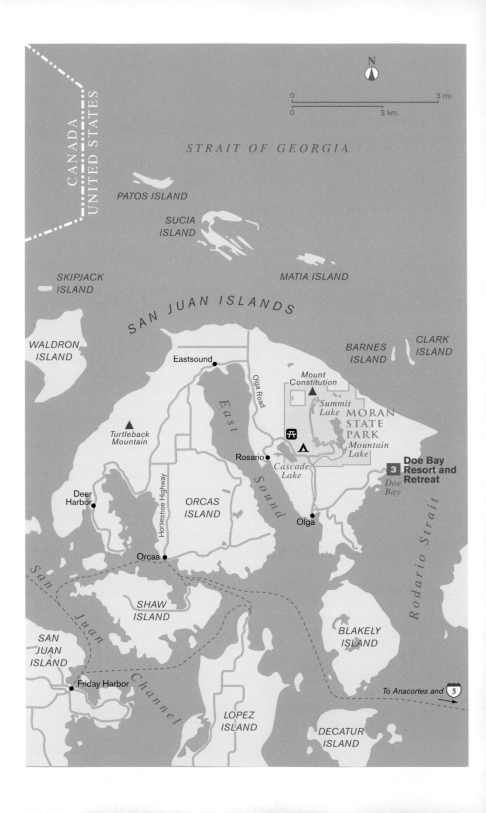

N

0 _____ 3 mi.
0 _____ 3 km.

CANADA
UNITED STATES

STRAIT OF GEORGIA

PATOS ISLAND

SUCIA ISLAND

SKIPJACK ISLAND

MATIA ISLAND

SAN JUAN ISLANDS

WALDRON ISLAND

BARNES ISLAND

CLARK ISLAND

Eastsound

Mount Constitution

Summit Lake

MORAN STATE PARK

Mountain Lake

Olga Road

East Sound

Turtleback Mountain

Rosario

Cascade Lake

3 **Doe Bay Resort and Retreat**

Doe Bay

Deer Harbor

Horseshoe Highway

ORCAS ISLAND

Olga

Rodario Strait

Orcas

SHAW ISLAND

BLAKELY ISLAND

San Juan Channel

SAN JUAN ISLAND

Friday Harbor

To Anacortes and **5**

LOPEZ ISLAND

DECATUR ISLAND

THE HOT SPRINGS

Years ago the brochure given to guests contained a line that still sums up the culture of this laid-back retreat: "Contrary to popular belief, Doe Bay Resort does not require guests to don tie-dye garb." Visitors describe Doe Bay Resort as a "neo-hippie" island retreat, with its roots extending back to the 1960s and 1970s, when a "human potential center" named the Polarity Institute was based on the property. The current managers warn that "Doe Bay is not for everyone, yet if you enjoy simplicity, nature, and relaxing, Doe Bay Resort is the place for you."

The resort welcomes guests of all persuasions, although you'll discover a strong liberal leaning in the politics of most. You may find yourself soaking in the hot tubs next to newlyweds on their first visit to Orcas Island or next to a Doe Bay veteran who has visited the resort every summer for the past twenty years to meditate and recharge her batteries.

The three mineral-water soaking tubs are reached along a footpath that skirts Otter Cove. Located on the opposite side of the cove from the Doe Bay Café, the tubs are sheltered by trees as well as by a translucent plastic roof that often drips with condensation from morning fogs. The three side-by-side soaking pools are each about 6 feet square and 3 feet deep. Two of the tubs are filled with mineral water that has

The mineral-water soaking tubs at Doe Bay Resort JEN EDINGTON, DOE BAY RESORT AND RETREAT

been heated to 105 to 110 degrees F, whereas the third, a cold-water pool, provides a respite from the thermal soaks. A spacious three-level sauna that can easily hold a dozen bodies is just a few steps from the pool. The pool and sauna area is often lively with conversation and laughter, especially in the evenings, but the atmosphere never gets too rowdy.

In between soaks you can enjoy excellent vegetarian and seafood meals at the Doe Bay Café. With lava lamps in the windows and Grateful Dead music playing in the background, the cafe reflects the true counterculture spirit of the resort. Breakfast features daily specials as well as a continental buffet; dinner usually has one or two seafood specials, as well as elegant vegetarian fare (beer and wine are also available). Pick a window seat with a good view of Otter Cove—during your meal the college-age waiter may point out a family of sea otters frolicking in the calm waters below your window.

History

The land around Doe Bay Resort was at one time the site of a fishing village for the Lummi tribe of Native Americans. The first Europeans homesteaded near Doe Bay in 1871, and the first post office was opened in 1881. In 1908 a new post office and general store were built near Otter Cove. This building, which is on the National Register of Historic Places, is now home to the Doe Bay Café. For decades the store and post office bustled with local fishermen and travelers arriving and departing from a nearby ferry dock. In 1953 the post office at Doe Bay closed. Since then the property has been the site of an artist colony, the "human potential center," and a health spa. The current resort is infused with the spirit of all these past activities. Joe and Maureen Brotherton have owned and managed Doe Bay Resort and Retreat since 2002.

WASHINGTON CASCADES

4. BAKER HOT SPRING

General description: An easily accessible soaking pool in a forest glen surrounded by cedars and Douglas fir. Trash and debris in and around the hot springs has been a problem in recent years, which detracts from the natural beauty of the area.

Location: Washington Cascades, 55 miles northeast of Mount Vernon in the Mount Baker–Snoqualmie National Forest.

Development: Undeveloped, except for the rock work done by volunteers to create the soaking pool.

Best time to visit: The hot spring is fairly empty during the mornings, mid-week, and wintertime. Avoid evenings and holidays unless you're looking to soak with a large (and sometimes rowdy) crowd. Weekends are popular with students from Western Washington University in nearby Bellingham, as well as with a few Canadians crossing the border to sample American hot springs.

Restrictions: The Forest Service's current policy is to "allow but not promote or encourage" the use of Baker Hot Spring. The agency also claims that the slow-moving water in the soaking pool often fails to meet water-quality standards for fecal-coliform levels. Bathe at your own risk. Unlike several hot springs on public land in Oregon and Washington, Baker Hot Spring has no nighttime soaking restrictions. The Forest Service has responded to a number of incidents of vandalism to cars in the parking area, broken beer bottles in the soaking pool, and obnoxious behavior by fellow bathers. Let's hope the inconsiderate actions of a small minority of visitors won't force the Forest Service to impose additional restrictions on the use of this pleasant soaking spot. You'll usually find a mix of nude and clothed bathers at Baker Hot Spring, so let your own judgment guide you on proper soaking attire.

Access: High-clearance vehicles are recommended to drive the last few miles if you take FR 1130 (which comes closest to the hot springs). Other drivers can park at the Park Creek Campground and hike the 3 miles to the springs. In heavy snow years the hiking may be difficult. Winter soaking enthusiasts sometimes bring their snowshoes or cross-country skis to traverse the final leg of the trip during snowy winters.

Water temperature: The pool temperature averages 100 to 105 degrees F, although it is sometimes cooler during spring snowmelt, when cold water from an adjacent stream mixes with the hot springs.

Nearby attractions: The overlook for Rainbow Falls is about 0.5 mile north of Baker Hot Spring on old FR 1144. This waterfall cascades more than 100 feet, forming a brilliant rainbow when the sun hits it just right. A nice nearby day hike is the Shadow of the Sentinels Trail. This gentle trail, located at milepost 14 on Baker Lake Road, features interpretive signs on the ecology and history of the area.

Services: None available at the hot spring. Food and gas are available year-round in the towns of Hamilton and Concrete on WA 20.

Camping: Park Creek Campground is situated near the intersection of FR 1144 and Baker Lake Road (about 3 miles from Baker Hot Spring). Boulder Creek Campground and Panorama Point Campground are also nearby. Swift Creek Campground (formerly Baker Lake Resort) is located about halfway up Baker Lake across the road from Park Creek. Swift Creek Campground has 50 private campsites that accommodate either tents or RVs.

Maps: Washington State Highway Map; Mount Baker–Snoqualmie National Forest map; USGS Mount Shuksan, WA; *DeLorme: Washington Atlas & Gazetteer,* page 17, C7.

GPS coordinates: N48.7639' / W121.6711'

Contact info: Mt. Baker–Snoqualmie National Forest, 2930 Wetmore Ave., Ste. 3A, Everett, WA 98201; (425) 783-6000 or (800) 627-0062; www.fs.usda .gov/mbs/.

Finding the springs: From I-5 north of Mount Vernon, take exit 230 east to WA 20. Drive 23 miles east on WA 20, then turn left at milepost 82 onto Baker Lake Road. Drive 20.5 miles up Baker Lake Road to the intersection of Swift Creek Campground and FR 1144 (on the left). Drive up FR 1144 to Park Creek Campground (less than 0.5 mile from the intersection). From Park Creek Campground FR 1144 has been decommissioned and is closed to all motorized use. It is a 3-mile walk from the campground to the hot springs on old FR 1144, or you can drive on FR 1130, which will put you much closer to the hot springs. To take the FR 1130 route, drive 18 miles up Baker Lake Road. Turn left onto FR 1130 (Marten Lake Road) just past the Boulder Creek Bridge. Drive 1.5 miles to a junction and keep right. Continue 2.5 miles from the junction to old FR 1144. Park at the junction and walk 0.5 mile on FR 1144 to the turnout and overlook. Look for the unmarked hiking trail to Baker Hot Spring on the right.

THE HOT SPRINGS

After being jostled on the bumpy forest roads, you'll savor the hushed quiet of the footpath that passes through a grove of old-growth cedar trees. The final few hundred yards of hiking ends in a cathedral of Douglas fir, maple, and cedar that surrounds the 15-foot-diameter soaking pool.

Baker Hot Spring can easily hold a dozen or more bathers, and unless it's really crowded you can stretch out and even paddle around a bit in the 2-foot-deep water. The hot springs water bubbles up in several places in the gravel bottom of the soaking pool. The pool temperature is about 104 degrees F at its hottest, but cools down slightly away from the bubbling inlets. There's a strong odor of sulfur in the water, and the sulfur smell will linger on your skin and hair long after you return to your car.

If left undisturbed, the water in the pool is crystal clear, but usually bathers stir up the sediments on the pool bottom. The water then takes on a unique bluish-gray milky color.

Candle wax on the rocks and logs surrounding the soaking pool testifies to nighttime gatherings, some of which can get pretty wild. You'll often find some trash and a

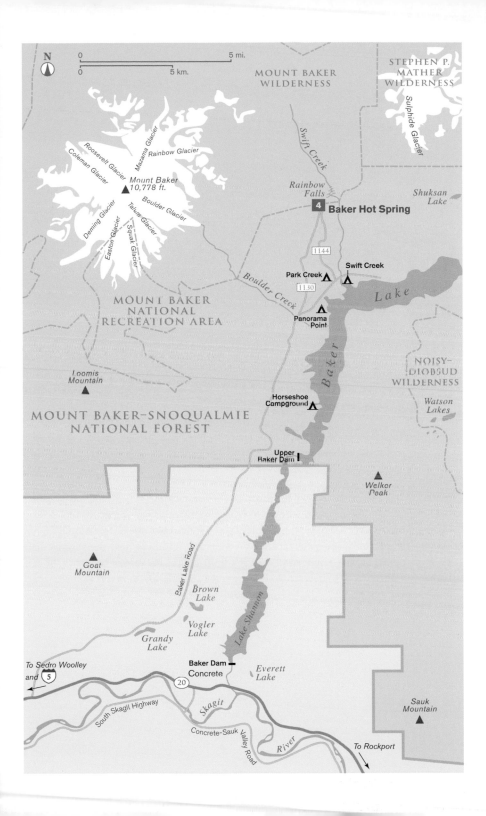

The Soaking Pool at Baker Hot Springs WAYNE ESTES

discarded towel or two along the trail or near the pool. Take along a garbage sack and pack out any trash to help keep this well-visited hot springs worth visiting.

History

In the 1890s a prospector named Joe Morovitz built a soaking pool around the natural hot springs. The springs were known as Morovitz Hot Springs for many years. No further development occurred until the 1960s or 1970s, when the Skagit Alpine Club built a four-person soaking tub, changing room, and outhouse near the springs. In 1978 the Forest Service removed all man-made structures because of concerns about high levels of coliform bacteria in the soaking tubs and a desire to revert the area to a more natural setting.

5. SCENIC HOT SPRINGS

General description: A mountainside soaking pool on private land surrounded by forest.

Location: Washington Cascades, 8 miles west of Stevens Pass Ski Resort, 61 miles east of Everett.

Development: A soaking pool has been built on the steep side of a mountain to collect the flow of a natural hot springs. In past years a variety of tubs and pools were built near the hot springs by volunteers, but the current concrete soaking pool was built under the direction of the private landowner.

Best time to visit: Although the hiking is easier in summer, a winter visit may be better for soaking since the soaking pools may be a bit cooler (and more comfortable).

Restrictions: All visitors to Scenic Hot Springs must first make a reservation through the Scenic Hot Springs website. A maximum of 10 soakers are allowed per day. Reserve your spot several days or more in advance. No soaking is allowed after sunset. A nominal fee per person is charged on weekdays, the fee is higher on weekends. Trespassing on the trail or at the pool without a permit can lead to a hefty fine. Nudity is the norm in the pools.

Access: The parking area at the trailhead just off US 2 may be blocked by snow in the winter. The 2.5-mile hike to reach the hot springs from the highway is difficult. Wintertime access may necessitate snowshoes and perhaps even crampons to navigate the snow and ice on the steep trail. Allow at least 2 hours to hike up the trail and an hour or so for the hike back to your car after soaking. Make sure you leave the hot springs early enough to get back before dark. Always check the Scenic Hot Springs website for the latest conditions and preparation advice before your trip.

Water temperature: The hot springs flow out of the mountainside at 122 degrees F. The temperatures in the three soaking tubs vary from 100 degrees F on the cooler side to 108 degrees F or more on the hotter side.

Nearby attractions: Stevens Pass Ski Resort and Iron Goat Hiking Trail.

Services: None at the hot springs. Bring food and water with you. Be prepared for hazardous weather in winter.

Camping: None at the hot springs. Two Forest Service campgrounds are within 10 miles of the trailhead—Money Creek and Beckler River. Detailed information on other nearby camping sites is available on the Scenic Hot Springs website.

Maps: Washington State Highway Map; *DeLorme: Washington Atlas & Gazetteer,* page 48, C1.

GPS coordinates: N47.7089' / W121.1381'

Contact info: http://scenichotsprings .blogspot.com. (You need to go to this website to make your reservations online.)

Finding the springs: Once you've made your online reservation to soak at the springs, the owners will e-mail you specific directions to the private parking area and hiking trail, as well as updates on weather conditions (parking distances from the trailhead may vary, depending on snow plowing and highway road closures). The hot springs parking area is about 61 miles

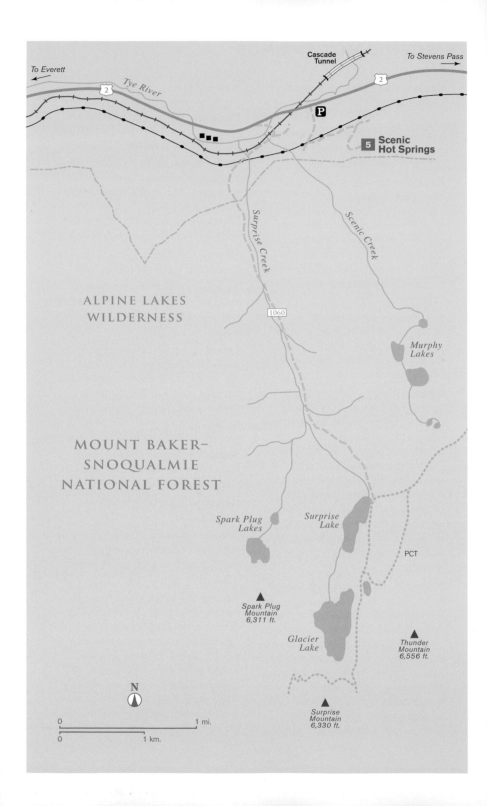

To Everett

Cascade
Tunnel

To Stevens Pass

Tye River

2

2

P

5 Scenic
Hot Springs

Surprise Creek

Scenic Creek

ALPINE LAKES
WILDERNESS

1060

Murphy
Lakes

MOUNT BAKER–
SNOQUALMIE
NATIONAL FOREST

Spark Plug
Lakes

Surprise
Lake

PCT

Spark Plug
Mountain
6,311 ft.

Thunder
Mountain
6,556 ft.

Glacier
Lake

N

Surprise
Mountain
6,330 ft.

0   1 mi.

0   1 km.

east of Everett on US 2. Parking your vehicle is the easy part—the hike to the hot springs is what you need to be physically prepared for. One visitor described the trail as "the steepest 2.5-mile hike ever!" The elevation gain during the hike is a bit over 1,000 feet. The first two-thirds of the hike follows a transmission line corridor (the corridor is private land, but the owner of Scenic Hot Springs has been granted access to allow permitted visitors to hike up the corridor to the hot springs). The final third of the hike takes an even steeper path before reaching the hot springs.

THE HOT SPRINGS

Once you have the hard hike to the hot springs behind you, the reward of three deep soaking tubs and a gorgeous view of the valley below make the effort worthwhile. The three oval tubs are made of black plastic, and hoses bring in hot water from the springs located above the tubs. Each tub can easily hold four to six people. The water in the tubs tends to be a little cooler in winter.

History

Rail workers building the first train route over Stevens Pass used the hot springs in the 1880s, but the hot water became most widely known when the Scenic Hot Springs Hotel was opened in the early 1900s. The hot springs water was piped down the mountainside for more than 2 miles to the hotel, where the water was used in mineral

Scenic Hot Springs RYAN COMMONS

baths for hotel guests. The hotel only survived until 1929, when rubble from a railway expansion was dumped on the hotel site. The hotel owners were granted ownership of the mountainside hot springs and some surrounding property as compensation for losing their resort to the railroad construction.

The train tracks a few miles west of Scenic Hot Springs were the site of one of the worst train disasters in the nation, when almost one hundred passengers were killed in March 1910 when their train was swept off its tracks by an avalanche.

6. GOLDMYER HOT SPRINGS

General description: A series of hot springs pools emerging from the mouth of an old mine, with a day limit on soakers to ensure a relaxing visit.

Location: Along the Middle Fork of the Snoqualmie River in the Cascade foothills, 60 miles east of Seattle.

Development: Amenities provided to visitors by Goldmyer are limited but include an open-air cabana at the hot springs pools, campsites with food-hanging lines and containers, 2 outhouses, 2 public picnic tables, and a bike rack.

Best time to visit: The hot springs are open year round, but access in the winter can be difficult due to road closures. Check the Goldmyer Hot Springs website for latest travel conditions and precautions.

Restrictions: Reservations are strongly encouraged (find more information on reservations on the Goldmyer website). Reservations can only be made over the phone at (206) 789-5631. It's best to make reservations at least 2 weeks in advance. Goldmyer Hot Springs is very popular, especially in the summer and on weekends, so book as far in advance as possible. Although it's not essential that you make advance reservations, the caretaker won't allow more than 20 people per day at the springs, so if you arrive without reservations and the quota is full, you'll be turned away. Nudity is the norm in the soaking pools at Goldmyer.

Access: The unpaved road to Goldmyer is challenging, with potholes, downed trees, and muddy ruts a common occurrence. It's best to have a high-clearance or off-road vehicle to make the trip. Road construction may also slow down your travel or even lead to a canceled reservation. Make sure to check the Goldmyer website for the latest information before starting your trip.

Water temperature: The hot springs emerge from the horizontal mine shaft at about 125 degrees F. The main pool at the mouth of the cave maintains a 110-degree-F temperature (on the high side for soaking), but the 2 overflow pools outside the cave mouth reach more comfortable temperatures, down to 104 degrees F.

Nearby attractions: Alpine Lakes Wilderness and Ollalie State Park.

Services: None at the hot springs. Bring all your own food and drinking water (or water purification kits).

Camping: Several camping spots are available at the hot springs. Since it will take most of a day to get to the hot springs, most visitors choose to camp overnight.

Maps: Washington State Highway Map; *DeLorme: Washington Atlas & Gazetteer,* page 47, F9.

GPS coordinates: N47.4856' / W121.3906'

Contact info: Northwest Wilderness Programs, 202 N. 85th St., #106, Seattle, WA 98103; (206) 789-5631; www.goldmyer.org.

Finding the springs: From Seattle drive east on I-90 to exit 34. Take exit 34, then turn left onto 468th Avenue. Continue on 468th Avenue for about half a mile to Middle Fork Snoqualmie Road, also known as FR 56. Continue on FR 56 for about 2 miles until the pavement runs out. Stay on FR 56 for another 11 miles, to the Middle Fork Trailhead parking lot. High-clearance

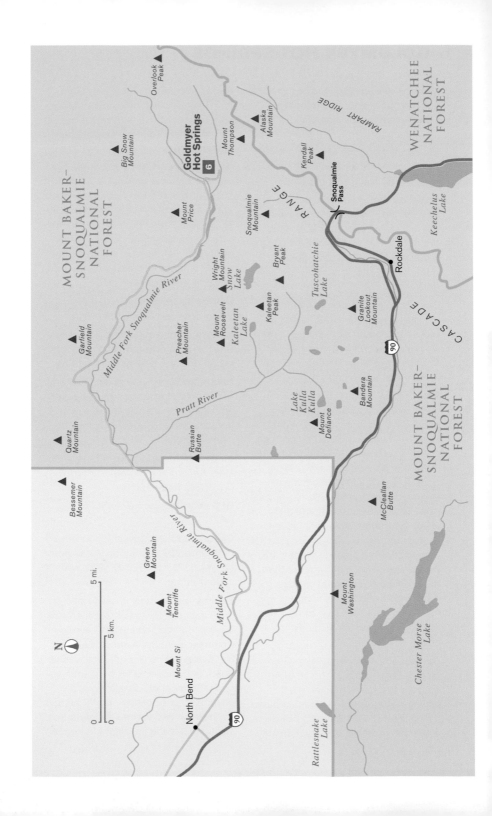

vehicles are recommended, since FR 56 is often muddy and potholed. Be sure to check the Goldmyer Hot Springs website for the latest updates on road conditions and construction.

Once you reach the Middle Fork Trailhead, park your car (and stash any valuables in your trunk if you aren't taking them with you). Follow the Middle Fork Trail for about 6 miles to the Dingford Creek Bridge. At the bridge you have 2 options. You can continue on the Middle Fork Trail for another 4 miles, or you can cross the bridge and follow Dingford Creek Road for 4 miles. The road option is a bit flatter and easier, while the Middle Fork Trail is more scenic, but you may run into more blowdowns and creek crossings. Both options will take you to the caretaker's cabin at Goldmyer Hot Springs. Ring the bell outside the cabin to register with the caretaker for your visit and instructions on camping and soaking.

Goldmyer Hot Springs WAYNE ESTES

THE HOT SPRINGS

Most hot springs emerge from the ground, but Goldmyer is unique—the source is 125-degree-F hot springs that flow from an abandoned horizontal mine shaft. A dam has been built at the front of the mine shaft, and the water pools behind this dam, forming a toasty soak that averages 110 degrees F. Overflow from the mine shaft pool cascades into two other cooler soaking pools. A cold-water pool is adjacent to the hotter pools, which lets you cool down between hot-water soaks. Since no more than twenty people are allowed per day at Goldmyer, you won't run into crowds, but you'll most likely be sharing the pools with other soakers.

History

A Californian named William Goldmyer first claimed and developed the hot springs in the early 1900s. Goldmyer built a small resort on the property, which was sold in the 1920s to Bill Morrow. A flood destroyed much of the resort in the 1960s, and the hot springs remained relatively unknown until the early 1970s when local media stories ignited a new interest in the hot springs.

Alarmed by the number of unsupervised visitors using the hot springs, the Morrow family created the nonprofit Northwest Wilderness Programs in 1976 to better preserve the area and control daily use. A caretaker's cabin was built near the hot springs in 1980, and a full-time employee of the nonprofit now lives on-site year-round.

7. CARSON HOT SPRINGS GOLF & SPA RESORT

General description: A once sleepy Victorian-era hotel and bathhouse that features a regimen of therapeutic hot mineral baths reminiscent of old European spas. The resort had a major renovation over the past decade and opened new guest and conference facilities, along with a magnificent golf course and restaurant/clubhouse.

Location: 50 miles east of Portland in the Columbia River Gorge.

Development: The hot springs have been developed commercially since the 1880s.

Best time to visit: Carson Hot Springs Golf & Spa Resort is open year-round. The number of visitors can increase tenfold on the weekends, so come midweek if you want more solitude. The popularity of the cabins and hotel rooms makes it advisable to book reservations well in advance.

Restrictions: A fee is charged to use the individual soaking tubs and other facilities in the bathhouse. Men and women use different restricted areas of the bathhouse, so most guests soak in the nude.

Access: Any vehicle can make the drive on the paved roads to the resort. Bathhouse hours vary, so check the resort website or give the office a call to check on current hours.

Water temperature: The hot springs emerge from the banks of the Wind River at 136 degrees F and cool to around 126 degrees F by the time they reach the bathhouse. The soaking tubs can be adjusted to any comfortable bathing temperature.

Nearby attractions: The Columbia Gorge Interpretive Center, near Stevenson, is a great place to spend an afternoon wandering through exhibits that detail the natural and cultural history along the Columbia River. Multnomah Falls, the second-highest year-round waterfall in the United States (and the most visited tourist site in Oregon), is located along I-84 just west of Cascade Locks on the Oregon side of Columbia River Gorge. Nine miles north of Carson on the Wind River Highway (Forest Highway 30) is the Forest Service's Wind River Information Center, which has lots of ideas for visiting area hiking trails and other waterfalls. The Upper Wind River Recreation Area is popular with cross-country skiers, snowmobilers and snowshoe aficionados.

Services: Elk Ridge Golf Course, just uphill from the hotel and lodge buildings, features a clubhouse with a restaurant serving breakfast and lunch. Lodging options are many at Carson, including simple Victorian-style rooms in the original Hotel St. Martin. A newer wing features an additional 28 rooms. The Wind River Wing features 39 rooms (22 of which have balconies with mineral-water hot tubs). Public baths, massages, and sweat wraps are available in the original bathhouse.

Camping: Camping and RV spots are available next to the lodge buildings. A number of other campgrounds are found both east and west of the resort along the Columbia River Gorge.

Maps: Washington State Highway Map; *DeLorme: Washington Atlas & Gazetteer,* page 100, C5.

GPS coordinates: N45.7288' / W121.7975'

Contact info: Carson Hot Springs Golf & Spa Resort, 372 St. Martin's Springs Rd., Carson, WA 98610; (509) 427-8296; www.carsonhotspringresort.com.

Finding the springs: From Portland drive east on I-84 along the Columbia River Gorge to Cascade Locks (exit 44). Turn north and drive across the Bridge of the Gods into Washington. Drive east on WA 14 for 3 miles to Stevenson. Proceed east of Stevenson for 3.3 miles on WA 14 and turn left at the Carson Junction. Drive 1 mile to the center of the small town of Carson and turn right at the 4-way stop onto Hot Springs Avenue. Drive 1 mile east on Hot Springs Avenue and turn left onto St. Martin's Springs Road, next to the entrance to the golf course. You'll see a sign here pointing toward Carson Hot Springs Golf & Spa Resort. St. Martin's Springs Road takes a sharp drop for 0.5 mile to the resort. Drive past the cabins and bathhouse to the old Hotel St. Martin and check in at the lobby.

THE HOT SPRINGS

Most visitors to this historic resort fall in love with the creaky Victorian-era hotel and the quirky hospital atmosphere of the bathhouse. With a prescribed, almost ritualistic approach, the resort follows the old European style of taking the mineral waters. New construction in the past decade has added modern lodging options without detracting from the charm of the existing resort.

The mineral baths in the century-old bathhouse are still a focal point of any visit to Carson Hot Springs. It's not a bad idea to make your reservation for a mineral-water treatment before you arrive. If you haven't done this, then schedule your mineral-bath session when you check in at the front desk in the hotel. You'll receive a pass to give to the white-uniformed bath attendants when it's time for your session.

The bathhouse was built in 1923 and looks it. The wooden building is divided into separate men's and women's areas, each painted an institutional white. Both sides contain showers and a dressing area, a tiled tub room, and a body wrap room. The bathroom attendants (male on the men's side, female on the women's) will ask you to disrobe and hang your clothes on a hook (leave your valuables in your car trunk or hotel room). You'll then be taken to the tub room, which contains two rows of claw-foot, cast-iron tubs that have probably hosted thousands of bare bodies over the decades. The attendant will fill a tub with a mixture of hot mineral water and cooler spring water, usually settling for a mixed temperature of 101 to 104 degrees F. (Individuals can increase or decrease the temperature as desired.) As you settle into the steamy water, the attendant will place a drinking cup under the hot-water tap and recommend you drink the sulfurous water while you soak. (Years ago bathers were exhorted to consume up to five cups of the unpleasant-tasting hot springs water during their half-hour soak, but at present drinking any of the water is up to you.) You'll then be left alone to soak in and imbibe the hot water for a half hour. The tubs are so wide and deep that it's hard to keep from floating, but you won't mind that sensation once the relaxing warmth of the mineral water starts to seep into your body.

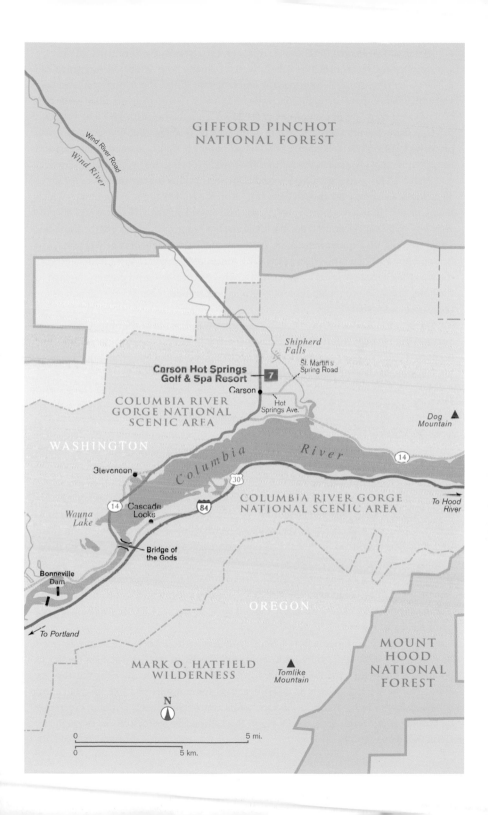

GIFFORD PINCHOT
NATIONAL FOREST

Wind River Road

Wind River

Shipherd
Falls

St. Martin's
Spring Road

**Carson Hot Springs
Golf & Spa Resort** 7

Carson

COLUMBIA RIVER
GORGE NATIONAL
SCENIC AREA

Hot
Springs Ave.

Dog
Mountain

WASHINGTON

Columbia   River

Stevenson

14

30

Wauna
Lake

14   Cascade
Locks

84

COLUMBIA RIVER GORGE
NATIONAL SCENIC AREA

To Hood
River

Bridge of
the Gods

Bonneville
Dam

OREGON

To Portland

MOUNT
HOOD
NATIONAL
FOREST

MARK O. HATFIELD
WILDERNESS

Tomlike
Mountain

N

0   5 mi.

0   5 km.

After you've soaked for about 30 minutes, the attendant will return and prepare you for what the resort brochure calls "a warm, relaxing body wrap." Most visitors call it "the sweat treatment." The attendant will help you out of the soaking tub and take you, still steaming and dripping, into the body wrap room, which is filled with two dozen "sweat cots." Before your body cools down from the hot bath, you'll be wrapped in a thin flannel sheet, then laid on one of the padded wooden cots. The attendant will place thick cotton blankets on top of you, tucking them tightly around your sides and feet, then wrap your neck and head in thick towels, leaving only your face exposed.

In the 1940s guests would lie on the sweat cots in this mummified state for up to 45 minutes. Current custom is to have you lie on the cot for about half an hour, which seems plenty. If you're claustrophobic, ask the attendant for a "loose wrap" to give your body a bit more room under the blankets, but longtime visitors ask for the "tight wrap" to build up a maximum sweat. No matter what the tightness of your wrap, in half an hour you'll be more than ready for a long, cold shower. Most guests whimper their way back to their hotel room or cabin, where they sleep until dinner, then eat in the hotel restaurant to regain their strength in preparation for another regimen in the bathhouse the following day.

In addition to the historic soaking treatments in the old bathhouse, Carson Hot Springs Resort now offers more modern spa treatments. A public mineral therapy pool is available to resort guests. Day visitors to the resort can obtain an hourly pass to use the mineral therapy pool as well.

History

Isador St. Martin discovered Carson Hot Springs in 1876 when he saw steam rising from the banks of the Wind River. St. Martin kept quiet about the existence of the hot springs until the land became available through the government, then he filed a claim to homestead the property.

St. Martin's wife had been suffering from a painful nerve disease called neuralgia, and St. Martin thought that she might benefit from bathing in the hot springs. He dug around the hot springs vent until he had made a hole large enough for his wife to bathe in. Mrs. St. Martin felt much better after these hot-water soaks, and soon word of the medicinal value of the hot springs spread to other people in the Columbia River Gorge.

An increasing number of visitors came to bathe in the thermal water, and St. Martin realized he could turn the hot springs into a commercial venture. At first he placed only a single soaking tub near the hot springs, but by 1900 he had built a large bathhouse near the river.

St. Martin envisioned a grand resort, complete with a hotel where guests could stay for a few days while they took therapeutic baths in the hot springs. He began building a three-story white clapboard hotel in 1897, completing the structure by 1900. The new hotel (which St. Martin named for himself) had twenty-four guest rooms, a large dining room, a kitchen, and a lobby. Supplies arrived at the resort by steamboat until the early 1900s when the railroad began operation through the Columbia River Gorge.

The Hotel St. Martin at Carson Hot Springs MARFA SCHERATSKI

In 1907 a small bathhouse was built near the hotel, supplied with water piped 275 feet uphill from the riverside hot springs. Cold water from wooden barrels was ladled into each tub to adjust the temperature to individual tastes.

Isador St. Martin was proud of his new resort and of the beneficial properties of the hot springs, but he had little toleration for visitors who were critical of the almost miraculous cures that he ascribed to the hot water. This characteristic may have indirectly led to his untimely death. An article in the *County Pioneer* from March 17, 1910, recounts how St. Martin met his tragic fate:

Fatal Affair at Carson

Isador St. Martin Stabbed by Robert Brown During Quarrel.

[A visitor to the hot springs] . . . had a quarrel about the quality of the water in St. Martin's Springs. This was always a sore spot with [Isador St. Martin], who, though he was of a quiet, industrious disposition, was always ready to scrap with anyone who made disparaging remarks about the quality of the water of these celebrated springs.

Old Man Brown made some disparaging remarks, which Mr. St. Martin wouldn't stand for, and had been ordered off the place. He was not going as fast as Mr. St. Martin thought he should, and Mr. St. Martin took him by the collar and started to push him along. When the old man struck backwards with a pocketknife, the blade penetrated just below the heart.

St. Martin died from the knife wound, but his heirs continued to run the resort for the next sixty-four years. Business flourished at the small resort, and in 1923 fourteen one-room cabins were built in a row above the road leading to the hotel. A larger bathhouse was constructed that same year, with separate wings for women and men. Massage rooms, soaking tubs, and rooms for sweat-cot treatments were included in the new bathhouse.

Local businessman Rudy Hegewald purchased the resort from St. Martin's heirs in 1974 and managed the hotel and other property for the next twenty years. In 1994 Hegewald sold the resort to Korean-based investor Gap Do Park for $3 million. The new owner announced his intentions to turn the sleepy resort into "the premier spa and resort in North America." Gap Park planned to invest $30 million to build a 200-room hotel, a 33,000-square-foot spa, an expanded golf course, and a 250-seat restaurant. Park's plans failed to materialize, and in 2011 Pete Cam, who owned the nearby Bonneville Hot Springs Resort & Spa, paid off the tax liens and purchased Carson Hot Springs Resort and the golf course from lenders. In 2013 Cam opened Carson's new lodging facilities and a revamped golf course to the public.

CENTRAL WASHINGTON

8. SOAP LAKE

General description: A small but vibrant spa town on the banks of a mineral-rich lake that has attracted health seekers since the late 1800s.

Location: Central Washington, 128 miles west of Spokane, 180 miles east of Seattle.

Development: The first homesteader arrived on the shores of Soap Lake in 1901. By the time the town of Soap Lake was incorporated in 1919, it already had 4 hotels and several rooming houses. Many of the elegant hotels and sanitariums are now gone, but the town has several hotels that cater to visitors from around the world who seek the benefits of the alkaline lake.

Best time to visit: The warm months of July, Aug, and Sept are best if you want to swim in Soap Lake and bake in the sunshine after taking a mud bath. The town hosts a popular Fourth of July celebration (check the Soap Lake Chamber of Commerce website for scheduled events). The cooler months of the year are much less crowded than the summer—bathing in the lake drops off dramatically after Labor Day. Winter visitors usually stay in one of the motels supplied with heated mineral water.

Restrictions: Swimsuits are required on Soap Lake's municipal beaches.

Access: Soap Lake is located on WA 17. Any vehicle can make the trip.

Water temperature: There's no thermal source heating the mineral water in Soap Lake, so the water temperature fluctuates with the seasons. The warmest lake temperature occurs in late summer, when the shallow lake can exceed 80 degrees F. The lake water is piped to several motels and other businesses in town. Many of the buildings in Soap Lake have dual water systems, one with freshwater and one with mineralized lake water. The mineral water is usually heated before it's used in showers and hot tubs.

Nearby attractions: Soap Lake is a great base from which to investigate many of the attractions of central Washington. A half-hour drive north of Soap Lake is the Grand Coulee Dam, the biggest concrete dam in the world. During evenings in July, Aug, and Sept, what's billed as "the world's largest laser light show" plays across the wide expanse of the dam (www.usbr .gov/pn/grandcoulee/visit/laser.html).

Between Soap Lake and Grand Coulee Dam is the Dry Falls Visitor Center, which explains the story behind what may have been the largest waterfall in history. More than 13,000 years ago, a gigantic glacial flood roared through central Washington. The overlook at Dry Falls provides a great view of the site where the waterfall, 400 feet tall and more than 3.5 miles wide, formed during the flood. Sun Lakes State Park, located a few miles from the Dry Falls Visitor Center, features camping sites, cabins, hiking, horseback riding, swimming, and golf.

Other nearby attractions include Lenore Caves, Summer Falls, Banks Lake, Steamboat Rock State Park, the Grant County Historical Museum, and the Gorge Outdoor Concert Amphitheater.

Services: Groceries, gas, food, and lodging are all available in Soap Lake. The town of Soap Lake has close to a dozen inns, motels, and resorts. Four of the most popular lodging options that feature mineral-water baths and showers are listed below:

Soap Lake Natural Spa and Resort (Formerly known as the Inn at Soap Lake and Notaras Lodge)
226 Main Ave. E.
Soap Lake, WA 98851
(509) 246-1132
https://soaplakeresort.com/

Masters Inn and Healing Retreat
404 4th Ave. NE
Soap Lake, WA 98851
(509) 246-1831
MastersInnSoaplake.com

Camping: The Smokiam RV Resort, located on Soap Lake's East Beach, features 36 RV sites and more than 50 tent campsites. A few of the RV sites feature private hot tubs. Cabins and tepees are also available. The larger Soap Lake RV Resort has 130 sites for tents and RVs.

Maps: Washington State Highway Map; *DeLorme: Washington Atlas & Gazetteer,* page 65, A7.

GPS coordinates: N47.3883' / W119.4875'

Contact info: Soap Lake Chamber of Commerce, 515 Main Ave. E., Soap Lake, WA 98851; (509) 246-1821; www.soaplakecoc.org.

Finding the springs: From Spokane drive 100 miles west on I-90 to exit 179 (Moses Lake/Ephrata exit). Drive north on WA 17 for 24 miles to Ephrata. Continue on WA 17 north past Ephrata for another 4 miles to Soap Lake. From Seattle drive 150 miles east on I-90 to exit 151. Take WA 283 north for 15 miles, until the highway turns into WA 28. Continue north on WA 28 for 11 miles to Ephrata. From Ephrata take WA 28 another 4 miles north to Soap Lake.

THE HOT SPRINGS

Soap Lake is a highly alkaline body of water that has no natural inlet. Coming from springs far beneath its surface, the lake's water passes through highly mineralized rock. The unique mineral component in the brackish water and bottom mud is the source of the healing properties sought by visitors to the little town on the lakeshore.

Summer bathers have their choice of two public beaches: East Beach is the most popular, with a snack bar, large beachfront, campground, and excellent views; bathers seeking solitude away from the city center gather at the more secluded West Beach.

Over the years several techniques have evolved for health seekers to use the mineral waters. A public fountain on Main Street allows visitors to fill gallon jugs with the mineral-rich water from Soap Lake. Some visitors return every summer to Soap Lake and pack their car trunks with dozens of gallon containers filled with the salty-tasting water.

Floating and swimming in the brackish water during summer is another popular way to enjoy Soap Lake. Brightly colored copepods (small shrimplike crustaceans) glitter in the sunshine in the lake, which is too salty to support any freshwater fish.

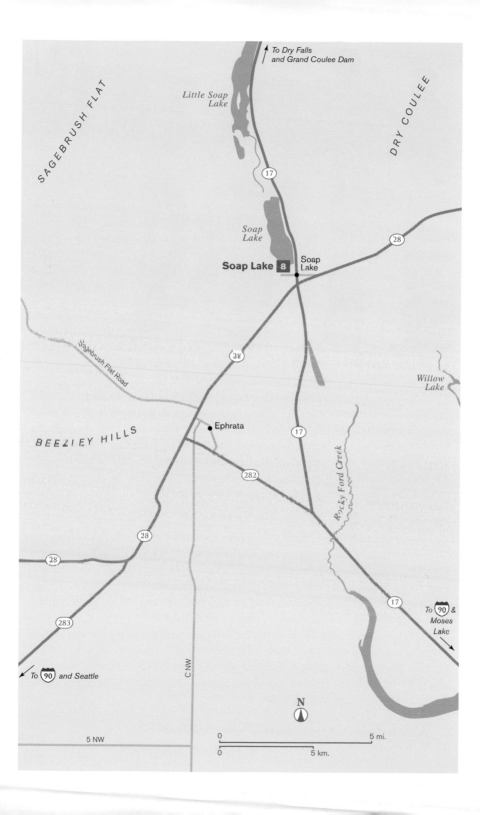

To Dry Falls
and Grand Coulee Dam

SAGEBRUSH FLAT

DRY COULEE

Little Soap
Lake

17

Soap
Lake

28

Soap Lake 8 Soap
Lake

Sagebrush Flat Road

28

Willow
Lake

Ephrata

17

BEEZIEY HILLS

282

Rocky Ford Creek

28

28

17

To 90 &
Moses
Lake

283

C NW

To 90 and Seattle

N

5 NW

0   5 mi.
0   5 km.

Enjoying the rejuvenating properties of the mud at Soap Lake KATHLEEN KIEFER

Mud baths are a third common method for benefiting from the minerals found in the lake. Soap Lake contains deposits of both olive-green and black mud, and locals treasure secret locations of their favorite mud, much like prospectors who hide the location of gold strikes. During the annual Fourth of July celebration, the lakeshore is lined with visitors who wait patiently to be coated with gooey mud by city volunteers. The slime-covered guests then bake in the sun until the mud dries to a crusty shell. A quick swim in the waters of Soap Lake then removes the mud.

History

In 1811 six Hudson's Bay Company trappers observed Native Americans from the Cayuse tribe taking steam baths in huts on the shores of Soap Lake. The Cayuse called the lake *Smokiam*, meaning "healing waters." They would heat a pile of rocks in a fire, then splash the rocks with lake water to create steam in the huts. Settlers in the late 1800s gave the lake its current name when they observed the frothy, soapy bubbles that whipped up around the lakeshore after a strong wind.

Like the Cayuse tribe before them, the early cowboys were aware of the healing properties of Soap Lake's water. Many tales were told of the miraculous cures realized by early pioneers who drank and bathed in the water. Some of these stories were pretty far-fetched. One cowboy legend concerning Soap Lake was described in the *Grant County Journal* on May 15, 1908:

> In the early days, a band of cowboys were camped near the lake, when one of their members died. His mates buried him near the shoreline and rode on down

the range another 50 miles. Two weeks later the boys were surprised to see their comrade (deceased) ride into the camp on a Cayuse [horse] he had picked up on the range. He said that he had been dead alright, but the water of the lake had seeped through until it had reached his body and brought him back to life.

Another story of Soap Lake's miraculous properties appeared in the same article:

Tony Richardson told us that a man was then stopping at the [Soap Lake] hotel who came 40 days before without hair on his head, and now his hair is two inches long.

A little over a decade later, the *Wenatchee Daily World* published the following two accounts of the healing properties of Soap Lake:

A well-known resident of Wenatchee, whose name can be furnished by inquiries, barely avoided an operation for appendicitis, which his physician insisted was the seat of his trouble. He instead came to Soap Lake, and after drinking the lake water expelled a tapeworm 36 feet in length and went home completely restored to health.

This story was also reported in the *Wenatchee Daily World:*

An old Swede from Spokane, 75 years old, was scabby all over with eczema, and was getting quite blind and had to quit work. He was put in the mud baths, and with only his nose sticking out, and commenced to sing, saying he was in heaven, and had not been free of pain for years until then. He was cured after taking 30 baths.

Rattlesnake bites were not uncommon in eastern Washington, and Soap Lake was known for its ability to treat victims of the venomous vipers. In May 1929 the *Grant County Journal* carried this letter to the editor about such a treatment:

To whom it may concern: After suffering from the after effects of a rattlesnake bite of a year and one half standing, in the form of stiffened muscles and joints, partial loss of use of one leg, blood destruction and heart trouble, and going everywhere for healing and health, the Mayo brothers advised a trip to Soap Lake as a possible place of relief. With the use of a cane I was able to slowly, and painfully, walk. After six weeks of Soap Lake treatment, I am able to walk as good as ever, and my other troubles are almost gone. There is no question in my mind but that Soap Lake water has curative powers that cannot be found elsewhere, and I offer this testimony from a fair and impartial view, as it looks to me.

Signed,
C.L. Nevins, DDS, LeMars, Iowa

Bathers leave their lounge chairs behind to slather on the mud at Soap Lake. KATHLEEN KIEFER

As the town became more cosmopolitan, some of the local customs had to be changed. A history produced by the Soap Lake Chamber of Commerce recalled one of these changes:

> Guns were finally outlawed in town because of the danger to swimmers caused by fun-seekers who sat on the shore and took pot shots at low flying birds over the water. The pistol packers were unhappy about being deprived of their evening entertainment, and argued that the birds were nuisances anyway.

Soap Lake flourished as a destination resort in the early 1900s, until several trage-dies struck in the 1920s when three of the largest hotels in town burned to the ground. The Great Depression in the early 1930s hit the tourist trade hard, and a long drought dealt a further blow to the economy of the area. The construction of the Grand Cou-lee Dam in the 1940s helped revitalize the spa town, although it never returned to its status as one of the most popular destination resorts in eastern Washington. Presently Soap Lake is a quiet outpost halfway between Seattle and Spokane. The twenty-first century has seen an increase in new residents, primarily retirees and persons seeking to escape the pressures of urban life.

OREGON HOT SPRINGS

A quiet pool at Breitenbush invites soakers to reflect and meditate. PETER PAUL RUBENS

OREGON CASCADES

9. BAGBY HOT SPRINGS

General description: A well-known backcountry soaking site nestled in a grove of old-growth cedar and Douglas fir. Hand-hewn log bathtubs and circular cedar group tubs filled with steaming hot water attract diverse crowds.

Location: Oregon Cascades, 65 miles southeast of Portland in Mount Hood National Forest.

Development: Over the past 40 years, the Forest Service and volunteers with the Friends of Bagby Hot Springs have constructed rustic bathhouses and hot tubs, but the surrounding old-growth forest still lends the area a wilderness feel.

Best time to visit: With downtown Portland less than a 2-hour drive from its trailhead, Bagby is one of the most popular hot springs in the Oregon Cascades. Weekends and holidays are particularly busy, and it's not uncommon to wait for an hour or so for your turn in the soaking tubs. Once the sun sets, Bagby can get rowdy—families might want to leave the area before dark. (The Forest Service now regularly patrols the area, and an attendant is usually on hand, which has calmed the party atmosphere somewhat.) You'll rarely find any time of the year when you'll have the hot springs to yourself. Try visiting in midweek in the morning or on rainy or colder winter days to beat the crowds.

Restrictions: No Northwest Forest Pass or day-use fee is required to park at the Bagby Trailhead or to hike up the trails. There is a nominal per person soaking fee (cash only;

pay the attendant at the trailhead) if you intend to soak. You'll be given a wristband that you need to wear while soaking. (You can also buy the wristbands with cash or credit card at the Ripplebrook store on the drive up to Bagby.) Nudity is permitted in the bathhouses but is discouraged in the open areas. No alcohol is allowed at the hot springs.

Access: Most of the year any vehicle can make the trip on the paved roads that lead to Bagby, although winter snows sometimes block the final few miles. After a snowstorm it's not unusual to see an abandoned car stuck in snow on the road to the Bagby Trailhead. (Call the Estacada Ranger District to check on road conditions in winter.) According to their brochure, the US Forest Service "strongly discourages visitors from attempting to drive or hike in once the roads are snow covered. These roads and trails are not maintained for winter travel and the area has no cell coverage for emergency calls. Visitors should not block travel ways with their vehicles when attempting to hike or snowshoe into areas. This makes it difficult for emergency services to access areas, and delays response times." In spite of this warning, backcountry skiers delight in visiting Bagby when the snows dissuade the usual crowds. Skiing distance from the road can exceed 10 miles one-way, depending on how far up you can drive on FR 70 before snowdrifts force you to park your car and strap on your skis.

Water temperature: The 3 hot springs that feed the hot tubs and bathhouses flow out of the ground at 136 degrees F. Cold water can be added to the tubs to cool them to a comfortable soaking temperature.

Nearby attractions: Kayakers and rafters enjoy the challenging water on the nearby Clackamas River. The Bagby Trailhead is a major gateway to the Bull of the Woods Wilderness. The Table Rock Wilderness, Salmon Huckleberry Wilderness, and the Olallie Lake Scenic Area are all located within an easy drive from Bagby.

Services: No services are available. Be sure to bring plenty of drinking water, as none is available at the hot springs.

Camping: To preserve the solitude of the hot springs, camping is not allowed near the soaking areas. Tent sites are available 0.25 mile beyond the hot springs at Shower Creek, along the Hot Springs Fork of the Collawash River. There's also an established campsite at the Bagby Trailhead. Backpackers stopping to soak at the springs sometimes continue another 7 miles south past the hot springs to Silver King Lake in the Bull of the Woods Wilderness. Another overnight option is Kingfisher Campground, which is maintained by the Forest Service, about 4 miles east of the Bagby parking area on FR 70.

Maps: Oregon State Highway Map; USFS Mount Hood National Forest map; *DeLorme: Oregon Atlas & Gazetteer,* page 35, A10.

GPS coordinates: N44.9357' / W122.1737'

Contact info: Clackamas River Ranger District—Estacada Ranger Station, 595 NW Industrial Way, Estacada, OR 97023; www.fs.usda.gov/

detail/mthood/recreation/?cid=fsb-dev3_053501. Also check out the website of the Northwest Forest Conservancy—Bagby Hot Springs, www.bagbyhotsprings.org.

Finding the springs: From Portland drive south on I-205 to exit 9 (the Estacada exit). Drive east on OR 224 for 18 miles to Estacada. Stop at the Estacada Ranger Station to get the latest information on road conditions and the hot springs trail. From Estacada drive 25 miles east on OR 224 to the Ripplebrook Ranger Station. About 0.5 mile east of Ripplebrook, turn south at the Timothy Lake Junction onto FR 46. Head south on FR 46 for 3 miles to FR 63. Take FR 63 south for 4 miles, then turn right (southwest) onto FR 70. Drive 6 miles southwest on FR 70 to the parking area for Bagby Hot Springs.

Another scenic way to get to Bagby is to drive east from Salem to Detroit along OR 22. From Detroit head north on paved FR 46, past Breitenbush Hot Springs and down the Clackamas River to the Two Rivers Picnic Area at the intersection with FR 63. Turn south onto FR 63, drive 6.1 miles to FR 70, and then turn southwest for 2.8 miles to the Bagby Hot Springs parking area. This route from Detroit is often closed in winter.

In past years the parking area at the Bagby Trailhead was infamous for vandalism, and car break-ins were common. Fortunately the Forest Service now has an on-site attendant who patrols the area, and problems with vandalism have dropped significantly. Even with the Forest Service oversight, it's best to leave your valuables at home, lock them in your trunk, or pack them with you to the hot springs.

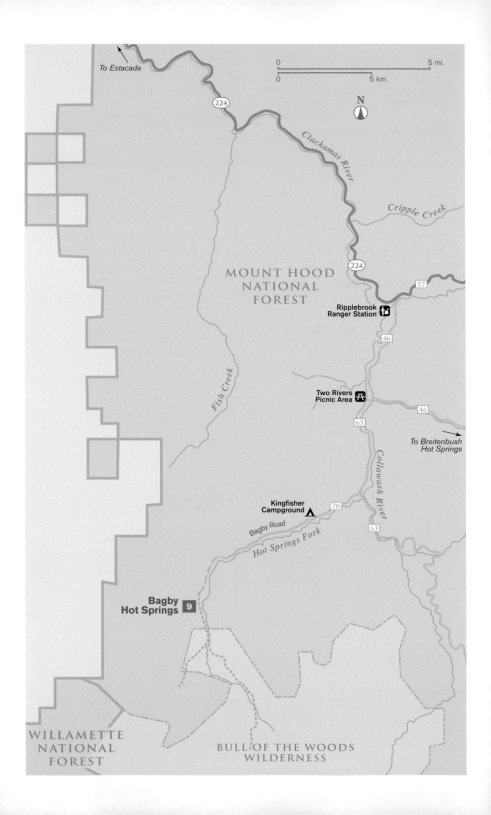

To Estacada

0   5 mi.

0   5 km.

N

224

Clackamas River

Cripple Creek

MOUNT HOOD
NATIONAL
FOREST

224

57

Ripplebrook
Ranger Station

46

Fish Creek

Two Rivers
Picnic Area

63

46

To Breitenbush
Hot Springs

Collawash River

Kingfisher
Campground

70

Bagby Road

63

Hot Springs Fork

Bagby
Hot Springs   9

WILLAMETTE
NATIONAL
FOREST

BULL OF THE WOODS
WILDERNESS

The trail to the hot springs is located just behind the information sign in the parking area. Allow 30 to 45 minutes to hike the 1.5 miles to the hot springs. For the most part it's a gentle hike on a well-maintained trail, but the last 0.25 mile is a bit steep. Pack a flashlight in your daypack—you'll need it to help you find your way back down the trail through the pitch-black, old-growth forest if you are hiking to your car after dusk.

THE HOT SPRINGS

Ask most hot springs veterans to name their favorite soaking spot in Oregon and Bagby is sure to come out near the top of their lists. The handcrafted rustic cedar bathhouses and soaking tubs seem designed more for hobbits and trolls than for the weekend millennial crowd from Portland.

There are three soaking areas surrounding the historic ranger cabin at Bagby. The community hot tub at the upper bathhouse is a spacious 6 feet in diameter and 4 feet deep. The tub holds six to eight people, and local custom is to invite newcomers to join those already in the hot tub. A wooden flume passes by the hot tub, carrying water from the 136-degree-F upper hot springs. A rubber ball the size of a grapefruit can be lowered into the flume, which diverts the hot-water stream into the hot tub (you'll figure it out once you see the flume). It takes about 30 minutes to fill the hot tub. There's also a valve that allows cold water from the nearby stream to mix with the steaming hot water in the tub. While you wait for the tub to fill with hot water, peruse the carved initials left by previous bathers on the sides of the hot tub.

A second 136-degree-F hot spring flows out of a rock fissure about 30 yards uphill from the two lower bathhouses. This hot spring flows for several yards through a channel cut in the surrounding rock, until it is collected by a second wooden flume. This flume is attached to the back of the lower bathhouses, on the opposite side of the bathhouse wall from the tubs. Until the summer of 2019, one bathhouse contained private tub rooms, each containing a hand-hewn, 10-by-3-foot tub made from a cedar log. Over the decades these hand-hewn logs rotted and became unsafe to use. They have been replaced with three oval soaking tubs.

A cedar roof covers half of each tub, leaving half exposed to the elements. If it's raining, you can orient yourself in the tub to keep your head covered under the roof, while your toes can wiggle in the raindrops.

The third bathhouse used to house three of the cedar tubs from Bagby's original bathhouse, as well as a large circular hot tub. The Forest Service installed new cedar tubs in this community bathhouse in 2011, replacing the old tubs. There aren't any private rooms in the community bathhouse, so if you want privacy, you'll need to wait for one of the private tub rooms to open up in the other bathhouse. Bathing suits are usually worn by soakers in the community tubs, but there may be a mix of clothed and unclothed soakers.

Between the two lower bathhouses lies a large wooden cistern that's filled with cold water from the nearby stream. Ten-gallon buckets are provided to haul water from the cistern to the tubs to cool them to a comfortable temperature (it usually takes several buckets to cool the water enough for bathing). After you've finished soaking at Bagby, be sure to use one of the large brushes located nearby to scrub down your tub for the next bather.

History

A prospector named Robert W. Bagby discovered the hot springs in 1881. Bagby apparently did little to develop the hot springs other than lend them his name. In 1913 a log cabin was built near the hot springs, which was used as a Forest Service ranger station through the 1940s. (The cabin is currently on the National Register of Historic Places.) In 1939 the Civilian Conservation Corps built the first bathhouse at Bagby. It featured five separate rooms, each containing a rustic bathtub carved out of a large cedar log.

Early visitors to the hot springs used mules to carry their gear on the three-day trek that followed winding forest trails. Because of its isolation, the hot springs received few visitors during the eighty years following Robert Bagby's first visit to the area. In 1960 the solitude was shattered when a new logging road allowed visitors to drive within a couple of miles of the hot springs. The popularity of the rustic location

Soaking tubs at Bagby Hot Springs

The outdoor soaking tub at Bagby Hot Springs CHRISTINE MERMILLIOD

boomed, and weekends often saw several hundred visitors to the springs. The Forest Service stationed a full-time ranger at the springs to oversee the crowds during these years.

In 1979 the original bathhouse was destroyed by fire. Many people suspected that a bather had left a candle burning in one of the tub rooms, which set the cedar-walled bathhouse ablaze. A local volunteer group called the Friends of Bagby Hot Springs was formed in the early 1980s to rebuild the bathhouse, and they also acquired a lease from the Forest Service to help manage the hot springs area. In 1983 this volunteer group built a hot tub near the upper hot springs and a year later constructed a bathhouse at the lower hot springs. Three of the original cedar tubs salvaged from the ashes of the original bathhouse were placed in the lower-springs bathhouse, along with a new hot tub. The Forest Service refitted the community bathhouse in 2011, replacing the original cedar tubs with three new, larger tubs, also made from cedar. The third bathhouse (still called the "New Bathhouse") was built adjacent to the lower-springs bathhouse in 1985 and was outfitted with five log tubs in private rooms.

10. BREITENBUSH HOT SPRINGS RETREAT AND CONFERENCE CENTER

General description: A worker-owned cooperative and intentional community situated in an old-growth forest, offering personal-growth retreats and group workshops. More than a half dozen hot springs pools and a natural sauna are available to retreat guests and day visitors.

Location: Oregon Cascades, 60 miles east of Salem.

Development: Breitenbush features rustic soaking pools in a mountain meadow, tiled soaking tubs near the Breitenbush River, and a log cabin sauna heated by natural hot springs. A central lodge and dining room, gift shop and registration office, meditation center, and guest cabins are located on the property.

Best time to visit: Workshops and retreats are held throughout the year, most of them available to the public (check the Breitenbush website or request a copy of their quarterly catalog for a list of future events). Breitenbush is busy during the week of summer solstice (late June), when several hundred visitors gather to celebrate shared community and spiritual renewal. The resort also hosts a popular Native American sweat lodge ceremony every month under the full moon—check your astrological chart (and the Breitenbush calendar online) if you're interested in this event. Certain weeks during the year are reserved for registered workshop participants only.

Restrictions: Day visitors are welcome to use the thermal pools and sauna from 9 a.m. to 6 p.m. (reservations required). Workshop participants and overnight guests can use the hot springs pools and sauna 24 hours a day. Nudity is the norm in the soaking pools and sauna, though the rest of the resort is not clothing optional. The retreat does not permit alcohol, electrical appliances, glass containers, candles, incense, or pets.

Access: Any vehicle can drive to Breitenbush on the paved and gravel roads leading to the retreat center.

Water temperature: The maximum measured surface temperature at the 5 dozen natural thermal springs is 198 degrees F. The meadow soaking pools are maintained between 100 and 105 degrees F. The 4 hot tubs near the river vary from 103 to 112 degrees F, and a fifth cold-water plunge rarely exceeds 60 degrees F.

Nearby attractions: More than 20 miles of hiking trails lead from the hot springs into the surrounding old-growth forest. During winter Breitenbush offers cross-country ski weekends into the Mount Jefferson Wilderness, about 10 miles from the resort. In the summer months Bagby Hot Springs can be reached by driving north on FR 46 through the Clackamas River Valley.

Services: Three vegetarian meals are served each day in the main lodge. Free well-being classes are offered 2 to 6 times daily, including yoga, meditation, and chanting. Trained therapists offer massage, hydrotherapy, and Reiki sessions. Lodging at Breitenbush is centered around 42 guest cabins, some with private toilets and sinks. A common bathhouse with separate men's and women's showers is available near the cabins. All cabins are geothermally heated. Bring your own sleeping bags, blankets, pillows, and towels.

Camping: Tent camping is available from May to Sept (with reservations). If you're on a budget and want cheaper overnight accommodations, consider camping at nearby Cleator Bend or Humbug Forest Service Campgrounds on FR 46. You can then pay a day-use fee to use the Breitenbush hot springs and sauna.

Maps: Oregon State Highway Map; *DeLorme: Oregon Atlas & Gazetteer,* page 36, C2.

GPS coordinates: N44.7831' / W121.9778'

Contact info: Breitenbush Hot Springs Retreat and Conference Center, PO Box 578, Detroit, OR 97342; (503) 854-3314; www.breitenbush .com; office@breitenbush.com.

Finding the springs: From Salem head east on OR 22 for 50 miles to the town of Detroit (at the edge of the Detroit reservoir). Turn north at the gas station in Detroit onto FR 46 and drive 10 miles. Just past the Cleator Bend Campground, turn right onto a single-lane bridge across the Breitenbush River. Drive 1.5 miles on this gravel road (FR 2231) to the Breitenbush parking lot. The parking lot was purposely built several hundred yards from the cabins, lodge, and hot springs to isolate the sounds of automobiles from the serenity of the resort. Handcarts are available in the parking lot to help accommodate the transport of your luggage. Register at the main office building to get directions to your cabin.

If you are driving from Portland, an alternate route to Breitenbush is to take OR 224 from Portland through Estacada to the Ripplebrook Guard Station and then continue on paved FR 46 to Breitenbush. This road is not maintained for winter travel and may be closed after the first winter snow.

THE HOT SPRINGS

Breitenbush Hot Springs Retreat and Conference Center is a great place to get away from urban pressures for a few days of reflection, relaxation, and renewal. The center is known in the Pacific Northwest for its emphasis on holistic living and personal wellness. Breitenbush encourages guests to be proactive in their own well-being and offers dozens of workshops and retreats throughout the year on such topics as conflict resolution in relationships, meditation, yoga, and rhythm and dance.

A sign near the main lodge claims that Breitenbush Hot Springs is the largest thermal springs area in the Oregon Cascades. Approximately sixty thermal springs and seeps occur on the hillsides and river valley near the resort, with an aggregate flow of 900 gallons per minute. The hottest thermal springs have a recorded temperature of 198 degrees F.

There are three separate areas where you can enjoy the thermal waters at Breitenbush. The meadow pools overlook the Breitenbush River, providing the most dramatic view of the surrounding mountains. At night these pools provide a wonderful opportunity for stargazing.

The biggest of the meadow pools is known as the "Sacred Pool," which measures 12 by 20 feet and is 2 to 3 feet deep. This pool often has a no-talking rule in effect to allow guests to soak in silence. Two smaller pools lie against the tree line between the

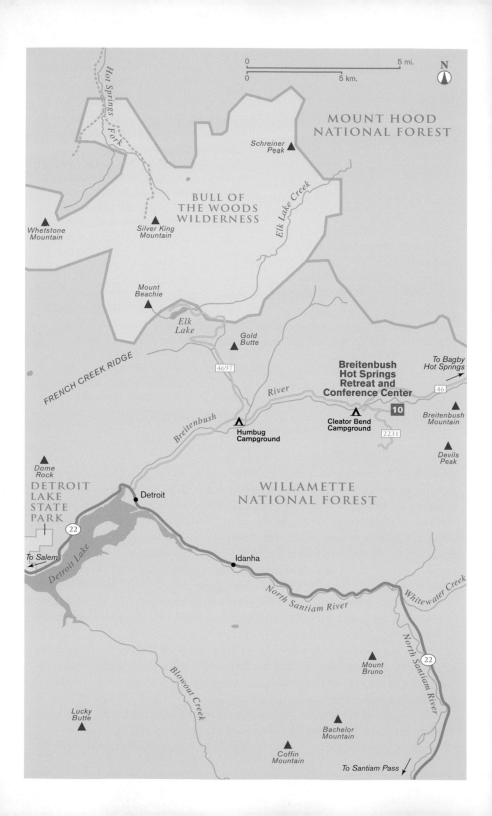

0   5 mi.

0   5 km.

N

Hot Springs Fork

MOUNT HOOD
NATIONAL FOREST

Schreiner
Peak

BULL OF
THE WOODS
WILDERNESS

Elk Lake Creek

Whetstone
Mountain

Silver King
Mountain

Mount
Beachie

Elk
Lake

Gold
Butte

To Bagby
Hot Springs

FRENCH CREEK RIDGE

4697

River

**Breitenbush
Hot Springs
Retreat and
Conference Center**

46

Breitenbush
Mountain

Breitenbush

Humbug
Campground

Cleator Bend
Campground

10

2231

Devils
Peak

Dome
Rock

DETROIT
LAKE
STATE
PARK

Detroit

WILLAMETTE
NATIONAL FOREST

22

To Salem

Detroit Lake

Idanha

North Santiam River

Whitewater Creek

North Santiam River

22

Mount
Bruno

Blowout Creek

Lucky
Butte

Bachelor
Mountain

Coffin
Mountain

To Santiam Pass

Quiet contemplation at Breitenbush Hot Springs BRINTON FOY REED

Sacred Pool and the main lodge. The temperature of the meadow pools hovers around 105 degrees F.

Adjacent to the main lodge is an old cedar steam sauna. A natural hot-water stream flows beneath the sauna, providing steamy heat to the dark wood interior of the building.

The Medicine Wheel Hot Tubs comprise the third soaking area at Breitenbush. The hot tubs are aligned with the four points of the compass and honor the four seasons. A cold-water tub provides welcome relief from long, hot soaks. The hot tubs are cleaned three times a day while guests are eating meals in the main lodge.

History

Breitenbush Hot Springs and the Breitenbush River were named for Peter Breitenbush, a one-armed Dutchman who settled at the mouth of the river in the 1840s. Hunters and trappers frequented the area through the 1800s. In July 1887 John Waldo of Salem visited the undeveloped hot springs. Waldo was suffering from an illness and reported that the mineral water had alleviated his condition:

> The water is helping me and could not well do otherwise if there is any virtue
> in it, for I sit for two hours of a morning snuffing it through a tube.

In 1893 Waldo again returned to the hot springs and reported on the new buildings constructed for persons seeking the medicinal benefits of the hot water:

One of many soaking options at Breitenbush Hot Springs BRINTON FOY REED

A log bath house, very plain, had been built over one of the hot springs, with a rather narrow cedar tub and a platform directly over the spring for a steam bath. Hot water is dipped out of the springs and cold water carried up from under the bank to temper it, and closing the door the steam from the spring makes a very agreeable chamber to bathe in. We all took a bath and pronounced it excellent.

In 1904 a settler named Mansfield applied for a homestead to the hot springs and surrounding land, selling it a few years later to a corporation. The corporation built tent platforms near the springs to provide sleeping quarters for summer visitors. In autumn the tents were removed from the platforms, and the area was abandoned until the deep winter snows melted the following spring.

Merle Bruckman purchased the hot springs in 1927. Bruckman's father had developed the technique for making ice-cream cones and had become wealthy after the new dessert was introduced at the St. Louis World's Fair in 1904. The 27-year-old Bruckman wanted to invest some of the family wealth in building a wilderness resort, and Breitenbush Hot Springs seemed perfect for his vision.

By 1930 Bruckman had built a 100-by-40-foot swimming pool, a bathhouse, guest cabins, and a hydroelectric generator. He began advertising the little resort as "Bruckman's Breitenbush Mineral Springs." Over the next few years, he added a gas station, a main lodge with thirty-two guest rooms, a grocery store, a post office, a soda fountain, a restaurant, and a dance hall. The resort flourished for two decades under Bruckman's management.

Nudity is common at the Breitenbush sauna. BRINTON FOY REED

In the mid-1950s Bruckman sold the property, and the resort soon began to decline. The property passed through a series of owners and finally closed to the public in 1972. For five years it sat abandoned, surrounded by barbed wire and "No Trespassing" signs.

In 1977, fifty years after young Merle Bruckman had first developed Breitenbush into a resort, another 27-year-old bought the property. Alex Beamer from Oakland, California, had been involved with self-sufficiency and spirituality for several years and was seeking a location to establish a personal-growth center. Beamer purchased the abandoned property for $250,000, using $100,000 of a family inheritance and borrowing the rest of the money.

Within three years Beamer and the volunteers who supported his vision had brought the little resort back up to code and had acquired a commercial business license. In 1985 Beamer sold his ownership to a newly formed cooperative that supported his ideals of a self-sufficient community built around personal growth. Since then Breitenbush has been managed as a worker-owned and worker-administrated cooperative corporation.

11. DEER CREEK (BIGELOW) HOT SPRINGS

General description: A warm soaking pool in a fern-lined grotto on the banks of the McKenzie River.

Location: Oregon Cascades, 61 miles east of Eugene in the McKenzie River Valley.

Development: Undeveloped, except for the crude dam of rocks that slows the influx of cold river water into the grotto.

Best time to visit: You are most likely to find a warm soak at Deer Creek Hot Springs when the river level in the McKenzie is low (late summer through winter). During spring runoff, cold river water can wash out the hot springs. Summer visits are preferable, as the soaking temperature is rather tepid. Weekdays are less crowded than weekends or holidays.

Restrictions: The hot springs are restricted to daytime use only. Swimsuits are optional.

Access: Any vehicle can make the trip up OR 126 and the paved forest road to the parking area.

Water temperature: A sluggish flow of hot water emerges from the grotto at around 130 degrees F, but the soaking pool itself averages 85 to 100 degrees F. The pool temperature can become quite chilly in spring, when the cold water from the McKenzie River washes over the crude dam of river rocks that enclose the soaking pool.

Nearby attractions: The McKenzie River Trail traverses the ridge a few yards behind Deer Creek Hot Springs. This hiking trail starts at Clear Lake and passes several scenic waterfalls as it parallels the McKenzie River for more than 30 miles. There are close to a dozen entrance points to the hiking trail along OR 126, and weekend visitors often hike short sections of the trail. Whitewater rafting and fishing are popular recreational activities on the river. Winter visitors enjoy cross-country skiing on the hiking trails. Downhill skiers gather at Hoodoo Ski Bowl on Santiam Pass, 20 miles east of Deer Creek Hot Springs.

Services: No services at the springs. Lodging is available at Belknap Hot Springs, 4.2 miles to the southeast. Belknap offers river-view rooms in its main lodge, as well as cabins with kitchen facilities, a camping area, and several dozen hookups for RVs.

Camping: Olallie Campground is located 1.5 miles northeast of the hot springs, just off OR 126.

Maps: Oregon State Highway Map; *DeLorme Oregon Atlas & Gazetteer,* page 43, C1.

GPS coordinates: N44.2411' / W122.0588'

Contact info: McKenzie River Ranger District, 57600 McKenzie Hwy., McKenzie Bridge, OR 97413; (541) 822-3381; www.fs.usda.gov/recarea/willamette/recarea/?recid=4210.

Finding the springs: From Eugene drive 53 miles east on OR 126 to the town of McKenzie Bridge. Proceed another 8 miles east on OR 126 and turn left onto FR 2654 (located between milepost 14 and 15). Drive on FR 2654 for 0.1 mile and cross the Deer Creek Bridge spanning the McKenzie River. Immediately after crossing the bridge, turn right into a small parking area. Lock your car and walk across the road toward the bridge. Both the popular McKenzie River Trail and a lesser-used footpath to Deer Creek Hot Springs parallel the McKenzie River. The McKenzie River Trail heads uphill—don't take this route. Instead look for the Deer Creek Hot Springs day-use sign next to the bridge. The trail you want is visible along the riverbank just behind the sign. Follow this trail downstream for about 100 yards to the hot springs.

THE HOT SPRINGS

Deer Creek Hot Springs is a mellow little soaking pool on the upper stretches of the McKenzie River. The hot springs are also known as Bigelow Hot Springs and McKenzie River Hot Springs, although the Forest Service has settled on Deer Creek Hot Springs as the official name. The hot springs are less popular than the larger (and hotter) soaking venues in the Oregon Cascades such as McCredie, Bagby, and Terwilliger, but they're still worth visiting if you're driving through the McKenzie River Valley.

There's only one soaking pool at Deer Creek Hot Springs, but it's a beauty. The 8-by-12-foot pool sits a few feet from the McKenzie River. Half of the pool is enclosed by a 4-foot-deep grotto covered with ferns and moss. You can choose to sit inside the

Autumn leaves surround Deer Creek Hot Springs. WAYNE ESTES

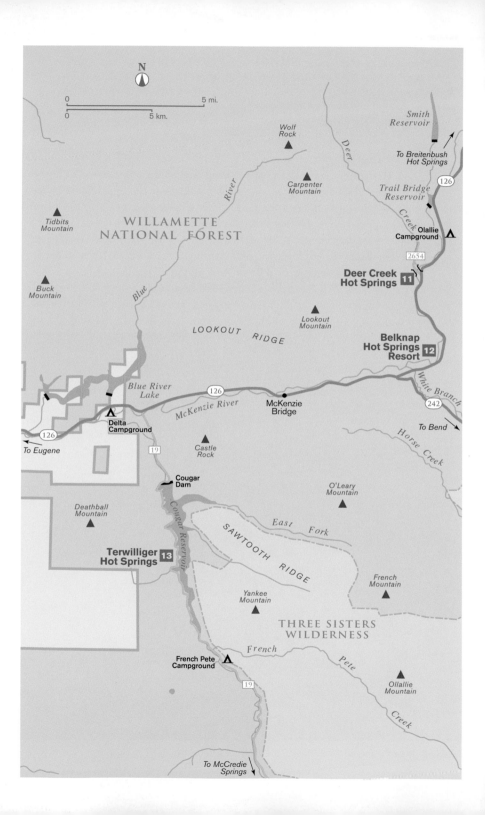

N

0 _____ 5 mi.
0 _____ 5 km.

Smith Reservoir

To Breitenbush Hot Springs

126

Trail Bridge Reservoir

Wolf Rock ▲

Carpenter Mountain ▲

Deer River

Olallie Campground △

2654

Deer Creek Hot Springs 🔟🔟

Tidbits Mountain ▲

WILLAMETTE NATIONAL FOREST

Blue River

Buck Mountain ▲

LOOKOUT RIDGE

Lookout Mountain ▲

Belknap Hot Springs Resort 1️⃣2️⃣

Blue River Lake

126

McKenzie River

McKenzie Bridge ●

White Branch

242

To Bend →

Delta Campground △

126
← To Eugene

19

Castle Rock ▲

Horse Creek

O'Leary Mountain ▲

Cougar Dam

Deathball Mountain ▲

Cougar Reservoir

SAWTOOTH RIDGE

East Fork

French Mountain ▲

Terwilliger Hot Springs 1️⃣3️⃣

THREE SISTERS WILDERNESS

Yankee Mountain ▲

French

French Pete Campground △

Pete

19

Ollallie Mountain ▲

Creek

To McCredie Springs ↓

A soaker enjoying Deer Creek Hot Springs STEPHEN HOSHAW

grotto (where the water tends to be a bit warmer) or soak in the exposed portion of the pool with a better view of the McKenzie River. Keep your eyes peeled for water ouzels bobbing and chirping on the rocks in the river current. These small brown birds actually fly underwater for short distances, searching for small insects. Rainwater and condensed mist from the hot springs collect in the ferns on the roof of the grotto, causing a gentle sprinkle of cold-water droplets on the bathers below. The bottom of the soaking pool is covered with a layer of fine silt, which muddies the pool water when disturbed. The soaking pool can comfortably hold four or more. When the river level is low, the pool can reach temperatures over 100 degrees F. Usually the pool temperature is a few degrees cooler, and you may find the pool washed out completely during spring runoff. The highway is barely visible through the trees on the opposite side of the river, but road warriors can't see the bathers. Nudity is common, though no one seems to care if his or her fellow bathers are wearing a swimsuit or soaking in the buff.

12. BELKNAP HOT SPRINGS RESORT

See map on page 58.

General description: A well-known commercial resort on the banks of the McKenzie River.

Location: Oregon Cascades, 55 miles east of Eugene.

Development: Belknap has been commercially developed for more than a century. Current facilities include 2 warm-water pools, a main lodge, rental cabins, and RV and camping spots.

Best time to visit: The resort is popular year-round. Rafting guides rendezvous with their clients in the Belknap Lodge before floating the McKenzie River in the spring and summer months; hunters pack the RV spots during the autumn deer season.

Restrictions: Swimsuits are required in the 2 resort pools.

Access: The paved highway and access road make Belknap Hot Springs Resort accessible by any vehicle. Pool hours are 9 a.m. to 9 p.m.

Water temperature: Located across the McKenzie River from the swimming pools, the near-boiling hot springs flow at a temperature of 196 degrees F. The 2 swimming pools are kept at 92 degrees F in summer and 102 degrees F in winter.

Nearby attractions: Belknap's location near the center of the McKenzie River Valley provides a great base for a variety of recreational activities. The 30-mile-long McKenzie River Trail starts at Clear Lake and passes several scenic waterfalls as it parallels the McKenzie River. Guests at Belknap can bike or hike on this trail, which passes within a few yards of the resort. A bit farther from Belknap, in the Tamolitch Valley, is a large sanctuary of old-growth Douglas fir trees. Whitewater rafting and fishing are within a few minutes' drive of the resort. Golfers can take advantage of the nearby Tokatee Golf Club, an 18-hole course. Cross-country skiing is available nearby, and the Hoodoo Ski Bowl is located 25 miles east of Belknap on Santiam Pass.

Services: The geothermally heated main lodge features 19 rooms, most of which overlook the McKenzie River and the lower swimming pool. Several of the lodge rooms feature their own hot tubs supplied with mineral water from the hot springs. All rooms in the main lodge are nonsmoking.

A dozen cabins are available on the upper terrace behind the main lodge, near the upper warm-water swimming pool. The cabins vary in size and can sleep 2 to 6. Forty-seven full-service RV spots are also available on the upper terrace and along the banks of the McKenzie River near the main lodge.

Cooking facilities and minifridges are provided in some of the cabins. Several restaurants are within 5 miles of the resort for visitors who want to stay at the resort and let others take care of their meals.

Camping: More than a dozen tent spots are available at the resort, and additional more-secluded spots are located across the McKenzie River. Several state parks are nearby for those who want to camp elsewhere and soak during the day at Belknap.

Maps: Oregon State Highway Map; *DeLorme: Oregon Atlas & Gazetteer,* page 42, D1.

GPS coordinates: N44.1906' / W122.0481'

Contact info: Belknap Hot Springs Resort, 59296 Belknap Springs Rd., PO Box 2001, McKenzie Bridge, OR 97413; (541) 822-3512; www.belknap hotsprings.com.

Finding the springs: From Eugene drive 53 miles east on OR 126 to the town of McKenzie Bridge. Continue 6 miles east of the town of McKenzie Bridge and turn left at the signs to the Belknap resort. Drive 0.5 mile to the main lodge building on the shores of the McKenzie River.

THE HOT SPRINGS

Belknap Hot Springs has been a popular destination resort for Oregonians for more than 150 years. Since the 1870s the scalding hot springs on the north bank of the McKenzie have been piped across the river to a variety of bathhouses, soaking tubs, and the two warm-water swimming pools. Day visitors can purchase an hourly or all-day pass to soak in the scenic lower pool.

East of the lower pool is a footbridge that leads across the McKenzie River to the source of the hot springs. Suspended from the bridge is the pipeline that brings the

The lower soaking pool at Belknap Hot Springs MARLENE WATSON

hot water from the springs to the resort. Past the hot springs are miles of hiking and biking trails that parallel the north side of the McKenzie River.

The upper terrace features several cabins and a second warm-water pool (which is only available to overnight guests of the resort). It is also home to the Belknap Gardens, several acres of formal landscaping that features both native and cultivated plants. Plan on taking a stroll on the walkways through the gardens after a relaxing swim in the riverside pool.

History

Belknap Hot Springs were discovered in 1854 by explorers searching for the headwaters of the McKenzie River. The hot springs were originally called "Siloam Springs" or "The Pools of Salome." No commercial use of the hot springs occurred until Rollin Simeon Belknap claimed them in 1870. Originally from Vermont, Belknap moved to San Francisco during the famous gold rush of 1849. A few years later Belknap moved to southern Oregon, where he fought in the Rouge River Indian War in 1855. In 1870 Belknap claimed the hot springs as well as some level ground on the opposite bank of the McKenzie River. In 1872 Belknap moved his family to the property and built a hotel and bathhouse on the river's south bank. The hot springs on the opposite side of the river from the little resort were piped through a flume made of split cedar logs that Belknap suspended across the river.

Belknap promoted his new resort to residents in the nearby city of Eugene. The following advertisement appeared in the *Oregon State Journal* in 1874:

To Those in Search of HEALTH OR PLEASURE

The undersigned, Proprietor of the SILOAM SPRINGS would call attention of those in search of health or pleasure to the properties and excellent situation of the above springs. They are situated on the McKenzie River, sixty miles east of Eugene City, surrounded by scenery beautiful and grand. The neighborhood abounds in game of every kind, and the streams with fine trout. The medicinal properties of the water have been tested by the cure of those who have visited them who have been afflicted with various disease, particularly Female Weakness, Scrofula, Rheumatism, Inflammations both external and internal, and general debility. Experienced males and females are always in attendance. Charges moderate. Good pasture nearby.

R.S. Belknap, MD

John W. Sims. Proprietor

Belknap's marketing proved successful, and a horse-drawn stagecoach brought visitors from Eugene to the resort several times a week. The stagecoach ride took 16 hours on bumpy dirt roads through the winding McKenzie River Valley.

The bridge crossing the McKenzie River at Belknap Hot Springs MARLENE WATSON

Siloam Springs became better known as Belknap Hot Springs. Belknap sold the property only five years after building the hotel and bathhouse, but the resort still carries his name.

Belknap Hot Springs passed through several owners over the next three decades. The ownership stabilized in 1907, when a wealthy lumber baron from Michigan named John Hawk acquired the property. Three generations of Hawk's family managed the resort over the next sixty-eight years.

The 1950s and 1960s were the heyday of the resort's popularity. Betty Smith, proprietor of Belknap Hot Springs during that time, recalled a typical summer: "Grandma and Pop came for the baths; Daddy fished and Mom knitted and the kids swam—it seemed like half of Eugene must have learned to swim up here."

Smith operated Belknap Hot Springs until 1967, when she closed the resort to the public. For the following eight years, the resort slumbered on the banks of the McKenzie River, with only a few of Smith's family members living on the property. In 1975 Smith sold Belknap Hot Springs to James Nation. Over the next three years, Nation rebuilt the neglected resort, reopening it to the public in 1978. Nation sold the property in 1995 to the McDougal family from Springfield, Oregon. The McDougals, who still own and manage Belknap Hot Springs, have refurbished the swimming pools, lodge, and restaurant and have expanded the manicured flower gardens on the upper terrace.

13. TERWILLIGER (COUGAR) HOT SPRINGS

See map on page 58.

General description: A cascading series of hot soaking pools in a valley of old-growth cedars and Douglas fir.

Location: Oregon Cascades, 53 miles east of Eugene in the Willamette National Forest.

Development: The hot springs have been captured behind stair-stepped rock and log dams, but they still retain a rustic wilderness feel.

Best time to visit: You will often have the pools to yourself early in the morning, especially in the wintertime and on drizzly days. Avoid late afternoons, weekends, and holidays if you want to miss the crowds from nearby Eugene.

Restrictions: Terwilliger is perhaps the most tightly regulated of all public hot springs in the Pacific Northwest. From the 1960s through the mid-1990s, the hot springs were embroiled in a debate over management strategies. More than 14,000 people visited the springs annually. Semi-permanent tent camps sprang up near the springs, with some visitors staying for weeks at a time. Violence and vandalism led the Forest Service to hold a series of public meetings in the 1990s to determine a better way to manage the hot springs.

In 1998 a day-use fee program was enacted, and camping was prohibited within 2 miles of the hot springs. Forest Service personnel began regular patrols of the area and imposed hefty fines on those who flouted the new rules. Fines of over $100 have been levied on soakers who were found in the hot springs after dark—so head back to your car before the sun sets.

Access: The hot springs are accessible year-round. Forest Service personnel drain and clean the soaking pools once a week, so you might want to call the Forest Service office and plan your visit for other days of the week. Occasional snows may block the access road for a few days in winter. It's an easy 10-minute hike on the gently sloping trail from the parking lot to the hot springs pools. Watch your step the last 10 yards when you cross the slippery rock dams that form the soaking pools. The pools are closed from 8 a.m. to noon on Thurs for cleaning.

Water temperature: The hot springs emerge from a small cave at about 116 degrees F. The stair-stepped soaking pools are hottest at the top, cascading downhill from a small cave at 112 degrees F to 85 degrees F in the bottom pool. The pools can be up to 5 degrees warmer in summer months and several degrees cooler in rainy weather.

Nearby attractions: Aufderheide Memorial Drive (FR 19) is one of the most scenic roads in the Oregon Cascades. The Blue River Ranger District office will lend you a free CD or audiocassette that gives you a mile-by-mile description of the natural wonders along this roadway. The 58-mile drive passes by Terwilliger Hot Springs, then climbs for 15 miles through old-growth Douglas fir to summit near Box Canyon Guard Station. Most day-trippers from Eugene turn around at the summit, but if you have extra time (and good brakes), continue on the winding descent for 32 miles to the towns of Oakridge and Westfir. McCredie and Wall Creek Hot Springs are only a few minutes' drive from Oakridge, so consider combining these popular soaks with your visit to Terwilliger to make a nice weekend getaway.

Services: No services are available at the hot springs except for composting toilets. Bring your own drinking water.

Camping: No camping is allowed within 2 miles of the trailhead to the hot springs. Several seasonal campgrounds are located south of the hot springs parking area. Drive along the west side of Cougar Reservoir on FR 19 for 3.5 miles to reach the French Pete Campground. The more secluded Homestead Campground is 10 miles south of the hot springs on the banks of the South Fork of the McKenzie River. Several other Forest Service campgrounds are available even farther south on FR 19.

If you're looking for a spot for your tent closer to the McKenzie River Valley, the Forest Service's Delta Campground is located near the Cougar Reservoir turnoff from OR 126 (7.5 miles north of the hot springs parking area).

Maps: Willamette National Forest map; DeLorme: Oregon Atlas & Gazetteer, page 41, C9.

GPS coordinates: N44.0836' / W122.2402'

Contact info: Willamette National Forest, 3106 Pierce Pkwy., Ste. D, Springfield, OR 97477; (541) 225 6300; www.fs.usda.gov/recarea/willamette/recreation/recarea/?recid=4391 (also visit the Friends of Cougar Hot Springs Facebook page: www.facebook.com/groups/101410912253/).

Finding the springs: From Eugene head east on OR 126 for 45 miles.

Between milepost 45 and 46 (about 4 miles past the town of Blue River), turn right onto Aufderheide Memorial Drive (FR 19). Drive 3.5 miles south on FR 19 until you reach Cougar Dam. Turn right at the dam and stay on FR 19. Drive 4.3 miles from the dam along the west side of Cougar Reservoir to the Terwilliger parking area. Park your car and purchase Terwilliger passes if you don't already have them (a per person day pass or an annual pass). The fee area includes the hot springs, trail, lagoon, and parking lot. You can pay on-site—bring cash, since credit cards may not be accepted.

Although there is significantly less vandalism in the parking area now than in past years, it's still wise to place all valuables in your car trunk or take them with you to the hot springs. Grab your towels and walk north along the side of FR 19 for about a hundred yards to the trailhead for the hot springs (the Cougar Reservoir will be on the right, and the lagoon and waterfall formed by Rider Creek will be on the left). The hike from the trailhead to the hot springs is less than 0.5 mile through a beautiful grove of cedars and Douglas fir.

A public-transit bus from Eugene stops at the Cougar Reservoir turnoff on OR 126, about 7.5 miles from the hot springs. On weekends college kids from the University of Oregon often ride this bus in the morning to the turnoff, then hitchhike or bike to the hot springs (the transit bus has bike racks). The transit bus picks up riders in the evening for the trip back to Eugene. Call (541) 687-5555 or check www.ltd.org for the daily bus schedule.

THE HOT SPRINGS

The popularity of Terwilliger Hot Springs is well deserved. The hushed stillness of the old-growth forest surrounding the hiking trail prepares visitors for the steamy soaks that await. The trail emerges near the uppermost hot pool, which, like all the pools, can hold five to ten people. Overflow from the upper pool forms a waterfall that drops 5 feet into the next pool. Three other pools follow in succession, each enclosed by a rock-and-log dam that keeps the pools between 1 and 3 feet deep. Water temperature ranges from a toasty 112 degrees F at the top pool to a cool 85 degrees F at the bottom pool. There's very little sulfur odor to the hot springs. The soaking pools rest on solid granite slabs, so there's little mud or debris to cloud the crystalline hot water. It's easy to find your preferred soaking temperature, because each pool is about 5 degrees cooler than the pool above it. The Friends of Cougar Hot Springs and the Forest Service have built rustic log-and-stone steps to aid navigation between the hot pools.

If you're visiting Terwilliger on a hot summer day in July or August, consider taking a dip in the nearby freshwater lagoon to cool off from the steamy waters of the springs. The lagoon is located at the base of the waterfall formed by Rider Creek.

One of several soaking pools at Terwilliger Hot Springs STEPHEN HOSHAW

Soaking pools at Terwilliger Hot Springs WAYNE ESTES

You can access the lagoon on several side trails leading off the main trail to the hot springs. All motorized craft are banned in the lagoon, so you can swim without worrying about Jet Skis or fishing boats passing nearby.

History

Terwilliger Hot Springs have been known by a variety of names. Capra Hot Springs appears to be one of the earliest monikers, followed by Rider Creek Hot Springs, South Fork Hot Springs, and, most recently, Cougar Hot Springs (so named because of the hot springs' proximity to Cougar Reservoir). Cougar Hot Springs is the name still used by many visitors, and a volunteer group called Friends of Cougar Hot Springs was formed in 1998 to help protect and maintain the soaking pools and hiking trail.

The Forest Service officially calls the area Terwilliger Hot Springs, named for Hiram Terwilliger, who came to Oregon in the 1860s and is thought to have been the first European to visit the hot springs. In 1906 Terwilliger filed a mineral-rights claim to the hot springs, but the Forest Service denied the claim, stating that Terwilliger's true intention was to build a summer resort on the property (a use not allowed under the mining law). Apparently the Forest Service changed its regulation two decades later, when in 1927 a special-use permit to construct a resort was issued to A. J. Jacobs of Eugene. Even with permit in hand, Jacobs never developed the property commercially, and the pristine hot springs remained little known.

Terwilliger Hot Springs soaking pool STEPHEN HOSHAW

For decades, visiting Terwilliger Hot Springs was a difficult proposition. Only the most committed travelers attempted the narrow trail along the South Fork of the McKenzie River. The isolated nature of Terwilliger Hot Springs abruptly changed in the 1960s when the US Army Corps of Engineers constructed a dam on the South Fork of the McKenzie to form Cougar Reservoir. A paved road was built from OR 126 along the west side of the new reservoir, which passed within a half mile of Terwilliger Hot Springs. Once the new road was finished, visitor use to Terwilliger skyrocketed.

From the 1960s through the 1990s, the hot springs increased in popularity. Visitors set up semi-permanent tent villages near the springs. At one time as many as 1,000 people were rumored to live in tents within a quarter mile of the hot springs. In 1997 members of the counterculture Rainbow Family scheduled their annual gathering near the springs, which attracted additional thousands of campers. Although most visitors during these years were peaceful, a few individuals caused problems. The Forest Service received an increasing number of reports of trash, illicit drug use, sexual assault, vandalism, panhandling, and noisy parties. The rowdy incidents peaked in 1996, when a man was killed by a neighboring camper who was angry at the late-night partying and loud music.

The increase in crowds and violence led the Forest Service to hold several public meetings to carve out a new management policy. In 1998 fees and restrictions on camping were imposed on the hot springs area, which quickly brought an end to the tent communities that had surrounded the hot springs. The fee system also provided funds that allowed the Forest Service to clean up the trash left by large crowds over the years, as well as rebuild the soaking pools. At present most visitors feel that Terwilliger Hot Springs is much cleaner and safer than it has been since the 1960s.

14. WALL CREEK HOT SPRINGS (MEDITATION POOL)

General description: A peaceful creek-side soaking pool surrounded by old-growth forest.

Location: Oregon Cascades, 51 miles southeast of Eugene.

Development: The pool around the natural warm springs has been enlarged to accommodate several bathers, and volunteers have built a crude rock dam to raise the soaking level. These small improvements don't detract from the primeval experience of soaking in an isolated glen of old-growth forest.

Best time to visit: Warm days in spring, summer, and autumn. Tepid water temperatures make this an uncomfortably chilly soak in winter.

Restrictions: A Northwest Forest Pass is required to visit Wall Creek Hot Springs. (Purchase the pass at the ranger station in Oakridge or Lowell prior to driving to the trailhead.) The trail and hot springs are day-use areas only.

Access: Any vehicle can make the easy drive on highway and black-top roads to the parking area at the trailhead. A gentle half-mile hike from the trailhead brings you to the warm springs pool.

Water temperature: The warm springs pool varies from 94 to 98 degrees F.

Nearby attractions: McCredie Hot Springs is situated about 10 miles southeast of Oakridge off OR 58. Consider a daylong hot springs tour, soaking at McCredie in the morning (before the crowds arrive), then visiting Wall Creek Hot Springs in the afternoon. Waldo Lake Wilderness and Diamond Peak Wilderness are about 20 miles east of Oakridge. Waldo Lake, Odell Lake, and Crescent Lake attract anglers and boaters.

Services: None available at the springs. Food and gas can be purchased in Oakridge, 10 miles southwest of the trailhead.

Camping: No camping is allowed along the trail or next to the hot springs. Salmon Creek Falls Campground, which is open from late Apr to mid-Oct, is located 5 miles southwest of the hot springs trailhead on FR 24. Motels are available in nearby Oakridge.

Maps: Willamette National Forest map, *DeLorme: Oregon Atlas & Gazetteer,* page 48, B3.

GPS coordinates: N43.8071' / W122.3111'

Contact info: Middle Fork Ranger District, 46375 Hwy 58, Westfir, OR 97492; (541) 782-2283; www.fs.usda.gov/recarea/willamette/recarea/?recid=4562.

Finding the springs: From Eugene drive south on I-5 for 3 miles to the Oakridge exit (OR 58, also called the Willamette Highway). Drive 40 miles east on OR 58 to Oakridge. Turn left off the highway at Oakridge's only stoplight, drive over the railroad track overpass, and turn right onto East First Street. Drive east on East First Street through Oakridge's main business district. At the east end of Oakridge, East First Street turns into FR 24 (it's a nicely paved 2-lane road). Drive on FR 24 for 9.1 miles on the winding road parallel to Salmon Creek. When you cross the bridge

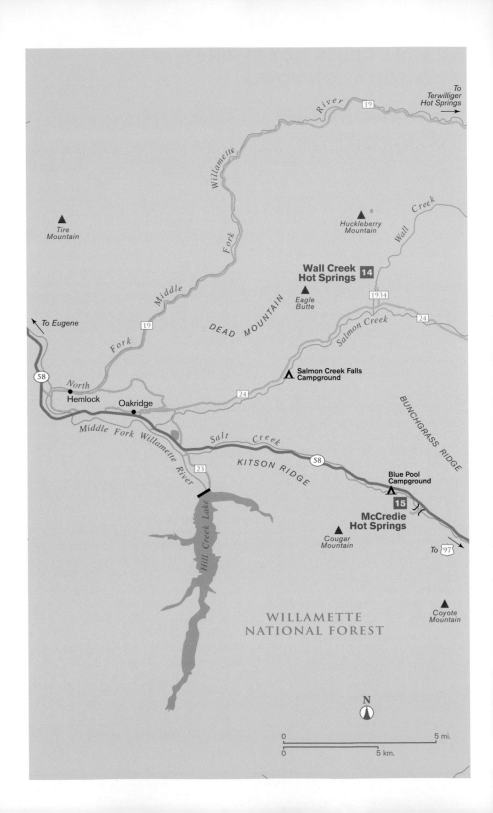

To
Terwilliger
Hot Springs

River 19

Tire
Mountain ▲

Willamette

Huckleberry
Mountain ▲

Wall

Creek

Fork

Wall Creek
Hot Springs 14

Middle

DEAD MOUNTAIN

Eagle
Butte ▲

1934

To Eugene

19

Salmon Creek

24

Fork

58

North

Salmon Creek Falls
Campground ⚠

Hemlock ●

Oakridge ●

24

BUNCHGRASS RIDGE

Middle Fork Willamette

Salt Creek

58

River

KITSON RIDGE

23

Blue Pool
Campground ⚠

15

Hill Creek Lake

McCredie
Hot Springs

Cougar
Mountain ▲

To 97

WILLAMETTE
NATIONAL FOREST

Coyote
Mountain ▲

N

| 0 | | 5 mi. |
| 0 | | 5 km. |

over Wall Creek, turn left onto FR 1934 at the Blair Lake turnoff. Drive 0.5 mile on FR 1934 (it's a dirt road). Turn left into the small parking area marked by a hiking trail sign. Place your Northwest Forest Pass on the dashboard and lock your valuables in your trunk.

The hiking trail from the parking area to the warm springs is a gentle 600-yard stroll. Douglas fir, hemlocks, maples, and moss-covered cedars line the trail. The hiking trail follows Wall Creek, ending in a small clearing that contains the warm springs pool.

THE HOT SPRINGS

Wall Creek Hot Springs slumbers on the banks of a fast-flowing stream. The soaking pool is about 10 feet wide and 3 feet deep, with perhaps half a dozen bubbling hot-water vents in the center of the pool. There's no noticeable sulfur odor to the water. The water temperature hovers around 95 degrees F—warm, but not hot enough for a good winter soak. The oval-shaped pool is about 12 feet wide by 15 feet long. Large river rocks encircle the pool's edge. The pool can easily accommodate five or six bathers. There's also a small hot-water seep (about 102 degrees F) 10 feet farther down the

Wall Creek Hot Springs (also called Meditation Pool) WAYNE ESTES

banks of Wall Creek from the main soaking pool. The lethargic flow rate of this seep (and the scum-covered rocks) makes this lower seep an unappetizing soak.

With the gentle sounds of nearby Wall Creek rushing by and the deep-green, moss-covered trees overhead, it's easy to see why Wall Creek Hot Springs is also called "Meditation Pool."

15. MCCREDIE HOT SPRINGS

See map on page 70.

General description: Hot soaking pools next to a stream just off an Oregon highway, frequented by locals, long-haul truckers, and tourists.

Location: Oregon Cascades, 51 miles southeast of Eugene.

Development: McCredie Hot Springs was a popular commercial resort until the 1960s, but all signs of the hotel and swimming pool are long gone. Volunteers have dug primitive soaking pools to capture outflow from the numerous hot springs along the banks of Salt Creek.

Best time to visit: Soaking at McCredie is popular year-round. Avoid holidays and weekends if you're seeking solitude. The secluded soaking pools on the south side of Salt Creek are less visited than the pools near the highway. The flow of water in Salt Creek can vary greatly during the year, from high floodwaters in the spring to an almost dry creek bed in the fall. This change in creek water flow can radically change the thermal soaking pools' locations and temperatures during the year.

Restrictions: The McCredie parking area and hot springs are located on Forest Service land and are closed from sunset to sunrise. Overnight camping is not permitted at the springs or in the adjacent parking area. A Northwest Forest Pass is not required. Nudity is allowed.

Access: Any vehicle can make the trip on OR 58 to the parking area. It's a 5-minute stroll from your car to the hot springs. Truck drivers often stop for a soak, as it's one of the few hot springs with a parking area large enough to handle the big 18-wheelers.

Water temperature: The hot springs at McCredie vary in temperature from 100 to 160 degrees F. Most of the soaking pools average a comfortable 100 to 105 degrees F. Be cautious when entering a new soaking pool at McCredie—the water temperatures can vary drastically from visit to visit.

Nearby attractions: Wall Creek Hot Springs is located about 10 miles northeast of Oakridge. A weekend trip could include visiting both Wall Creek Hot Springs and McCredie Hot Springs and staying in one of the many area campgrounds or at one of the motels in Oakridge. Great hiking opportunities are available in nearby Waldo Lake Wilderness and Diamond Peak Wilderness, about a half hour east of Oakridge. Fishing and boating are popular activities at Waldo Lake, Odell Lake, and Crescent Lake.

Services: No services are available at the hot springs. Restaurants, supplies, and gas can be found in Oakridge, 9.5 miles northwest of the hot springs on OR 58.

Camping: The Blue Pool Campground, maintained by the Forest Service, is located 0.5 mile west of the hot springs. The campground is open from mid-May to Oct. You can easily walk to the hot springs from the campground, so consider staying there a night or two and spend your daylight hours soaking at McCredie. Motels are available in nearby Oakridge.

Maps: Oregon State Highway Map; *DeLorme: Oregon Atlas & Gazetteer,* page 48, C4.

GPS coordinates: N43.7060' / W122.2885'

Contact info: Middle Fork Ranger District, 46375 Hwy. 58, Westfir, OR 97492; (541) 782-2283; www.fs.usda .gov/recarea/willamette/recreation/ recarea/?recid=81565.

Finding the springs: From Eugene drive south on I-5 for 3 miles to the Oakridge exit (OR 58, also called the Willamette Highway). Drive southeast for 40 miles to the town of Oakridge. From the stoplight in Oakridge, continue driving southeast on OR 58 for another 9.4 miles to the Forest Service's Blue Pool Campground. Continue another 0.5 mile on OR 58 past the campground to milepost 45. Look for a large parking lot just past milepost 45 on the south side of the road, near a sign that says McCredie Station Road. Pull into the parking lot. There is parking for 9 passenger vehicles and 1 40-foot RV. Two picnic tables are available. Overnight camping is not allowed. The trail to the hot springs is located at the east side of the parking lot. Lock your car, grab your towel, and walk about 50 yards along a dirt path that parallels Salt Creek to the soaking pools.

If the soaking pools on the north side of Salt Creek are filled with fellow hot springs enthusiasts, then give the hot springs on the other side of Salt Creek a try (although the south-bank pools may be too cold to enjoy for very long). To reach these hot springs, walk back to the main parking lot, then drive 0.5 mile southeast on OR 58. Turn south onto Shady Gap Road and cross the bridge spanning Salt Creek. Turn right (west) just after crossing the bridge onto FR 5875. Drive for 0.1 mile toward the first sharp left curve in the road. There's a "Day Use Only" sign barely visible in the trees to the right (toward Salt Creek). Park your car on the wide pullout just past the sign, then walk back a few yards from your car until you see a faint trail heading down the bank and downstream along Salt Creek. Follow this trail for about 500 yards (keep left at the 2 branches in the trail) until you reach the shaded soaking pools on the south bank of Salt Creek. You'll see the more heavily visited soaking pools across the creek about 100 yards to the north. (**Caution:** It's tempting to wade across Salt Creek from the north-bank soaking pools to the more secluded pools on the south bank, but the fast-moving current in the creek can be treacherous, so be sure the water level is low enough for a safe fording of the creek. If the water is high, take the extra 15 minutes to drive across the bridge and hike the trail to the south-bank soaking pools.)

THE HOT SPRINGS

The 50-yard walking path from the parking area by the highway will bring you to an open terrace on the north bank of Salt Creek, which contains a variety of soaking opportunities. Hemlocks, cedars, and Douglas fir separate the sunny, open stream bank from the highway, which lies 100 yards north of the soaking area. A car or truck on the highway may occasionally be seen through the trees, but for the most part the traffic has little impact on the soaking experience.

Pool depth, temperature, and location can vary from season to season along Salt Creek, but the following descriptions cover the usual locations and temperatures of the soaking pools.

The largest of McCredie's soaks on the north bank is a crowd pleaser that's about 20 feet wide by 40 feet long and 1 to 2 feet deep. Three separate hot springs on the edge of this pool keep the soaking temperature around a comfortable 105 degrees F (depending on how close you sit to a hot springs vent), and several smaller vents bubble in the pool itself. Some of the hot springs that supply the soaking pools at McCredie exceed a scalding 160 degrees F, so test the pools before you get in and avoid getting too close to the hot springs vents. There's a slight sulfur smell in all of the pools at McCredie, but it's quite tame compared with the smell in some other hot springs in the Cascades.

About 20 yards upstream from the main pool is a second sizable soaking opportunity—a 20-foot-diameter circular pool that can hold a modest (or immodest) crowd. This pool is slightly less popular than the large soaking pool.

Bordering Salt Creek are usually two or three rock pools suitable for one to two people. These small pools capture hot water from 150-degree-F springs that flow into the creek. Cold water from the creek mixes with the hot springs water in these pools, resulting in a comfortable 100-degree-F average temperature. These pools may be washed out during high water in the creek in the spring.

The more secluded hot springs on the south side of Salt Creek are also well worth checking out, especially if the north-side pools get too busy. The main source of geothermal heat on the south bank is a 125-degree-F spring issuing from the base of a low, lichen-covered support wall of river rock and concrete. This crumbling foundation wall is all that is left of the old McCredie Springs Lodge and swimming pool. The main spring flows into a pool about 10 feet in diameter and 1 foot deep. Temperature in this pool is usually a comfortable 100 to 105 degrees F. A rock dam separates this main pool from a slightly smaller (and slightly cooler) lower pool. These two pools lie in a shady grove of cedars, maples, and Douglas firs. It's certainly more secluded and peaceful in these pools than in those on the sunny north side of Salt Creek.

History

In 1878 a trapper named Frank Warner discovered the hot springs while following a Native American trail along Salt Creek. Warner built a cabin near the springs but apparently did no other development to the property. When the national forest system was established in the early 1900s, the hot springs were declared public property.

John Hardin, a builder from Eugene, filed a placer claim on the hot springs in 1911. (Under mining laws at that time, anyone who found salt deposits on public land could file a mining claim.) Hardin was given a mining lease on the property, but his real intention was to build a resort near the hot springs. By 1914 Hardin had completed construction of a two-story hotel that could house sixty guests. The hotel featured a large porch that extended along the entire south side of the hotel and overlooked Salt Creek. Large doors opened up onto the porch from each bedroom so that

A soaking pool at McCredie Hot Springs

a bed could be rolled out on the porch on hot summer days to take advantage of any cool breezes near the creek.

In 1916 the lease for the resort was transferred to Judge Walter McCredie, a colorful character from Portland who owned a semipro baseball team. The judge would often bring his team members to the resort to enjoy the benefits of the hot water. Although McCredie managed the resort for only five years, the hot springs are still associated with his name.

The Southern Pacific Railway, built through the valley in 1923, passed within a few yards of McCredie Hot Springs. This opened up the entire area to tourists from Eugene and Portland. During the resort's heyday in the late 1930s, five trains stopped daily near the resort.

The McCredie resort continued to operate through the 1940s and 1950s, although it suffered through a series of poor managers. The resort developed a particularly unsavory reputation in the late 1940s when a woman described by a Forest Service representative as "one of the most impossible persons we have ever known" managed the property. According to a cultural history produced by the Forest Service, the woman ran a bordello out of the resort, with three daily shifts of prostitutes—"one working, one coming, and one going." Although the resort's manager was repeatedly jailed for running a house of prostitution, she always managed to make bail and was "as slippery as an eel at evading the law." The deputy sheriff in nearby Oakridge stated that the woman had alone "caused his office more trouble than the entire population of Oakridge and Westfir, with fringe areas put together, some eight to ten thousand people."

Even though Forest Service officials were aware of the shenanigans at McCredie, they failed to shut down the bordello. The cultural history written by the Forest Service noted that "she had guns and boasted she knew how to use them. Forest Service officials seemed to prefer to remain healthy."

The feisty manager of the resort left peacefully in the 1950s, when George Owen acquired the lease. Owen's tenure at McCredie was filled with tragedy. Fire destroyed the hotel in 1958, and a flood on Christmas Day in 1964 washed out the bridge across Salt Creek and destroyed the swimming pool. The Forest Service terminated Owen's lease shortly after the flood and then burned or removed all remaining buildings on the property. At present only the concrete foundations of the old hotel are visible near the soaking pools on the south side of Salt Creek.

16. UMPQUA HOT SPRINGS

General description: A series of soaking pools atop a 100-foot terrace that overlooks the North Umpqua River.

Location: Oregon Cascades, 60 miles east of Roseburg, 30 miles west of Crater Lake National Park.

Development: The hot springs are undeveloped, except for the primitive 3-sided shelter that encloses the main soaking pool on top of the terrace.

Best time to visit: The Forest Service conducted a user survey of Umpqua Hot Springs in the late 1990s and found that close to 9,000 people visited the springs every year— and Umpqua's popularity has only increased since then. Try to visit in the early morning and during the middle of the week to avoid crowds. You may have to wait your turn to soak in the hot springs on holidays and week-ends. Cooler days in autumn, winter, and spring may be more enjoyable for soaking in the 106- to 115-degree-F pools. (There is no cold-water source to reduce the pool temperature, so a soak on a hot summer day may not be as enjoyable as it would be during cooler weather.)

Restrictions: A Northwest Forest Pass is required for all vehicles parked at the Umpqua Hot Springs parking area. No camping is allowed at the hot springs. No motorized vehicles are allowed on the access trail. Nudity is common in the soaking pools.

Access: The hot springs are open year-round, 24 hours a day. Any vehicle can make the trip on the paved and gravel roads to the Umpqua Hot Springs parking area, although occasional winter snows can block the last few miles of road. The 0.3-mile hiking trail is a bit steep in the last 200 yards before reaching the hot springs. The Forest Service closes a gate just past the junction of FR 34 and FR 3401 to prevent vehicles from getting stuck in the snow on the way to the hot springs. Hundreds of cars have gotten stuck trying to reach the Umpqua parking area in the winter, which led the Forest Service to start closing the gate during the snowy winter months. If the gate is closed, you'll have to hike, snowshoe, or ski an additional 2 miles to reach the hot springs.

Water temperature: The temperature in the soaking pools varies between 100 and 115 degrees F, with tempera-tures between 106 and 110 degrees F most common.

Nearby attractions: The Umpqua National Forest is home to an amazing variety of outdoor recreational oppor-tunities. Crater Lake National Park lies less than an hour from Umpqua Hot Springs. The North Umpqua River Trail features many access points along OR 138, making it perfect for day hikes or a weeklong trek. The North Umpqua River is popular with rafters and anglers pursuing steelhead salmon. Winter turns many area hiking trails into snow-covered paths popular with cross-country skiers, and a half-dozen "snoparks" between Toketee Lake and Crater Lake National Park give snowmobilers a variety of backcoun-try recreation opportunities.

Services: The nearest full-service town is Roseburg. Limited supplies can be purchased 12 miles away at the Dry Creek Store. No drinking water is available at the hot springs, so be sure to bring plenty with you.

Camping: A few unofficial camping spots are located across the North Umpqua River downstream from the hot springs. (You may be able to see some tents from the Umpqua Hot Springs soaking pools.) To reach these riverside camping spots, take the trail that forks downhill from near the composting toilet adjacent to the hot springs. Campers also often stay near their cars in the Umpqua Hot Springs parking area overnight, to be ready for an early morning hike to the hot springs to beat the crowds that can show up later in the day. There's another camping area at Toketee Campground, 2.5 miles southwest of the hot springs parking area at the head of the Toketee Reservoir.

Maps: Oregon State Highway Map; Umpqua National Forest Recreation Map; DeLorme. Oregon Atlas & Gazetteer, page 55, B8.

GPS coordinates: N43.2958' / W122.3659'

Contact info: Diamond Lake Ranger District, 2020 Toketee Ranger Station Rd., Idleyld Park, OR 97447; (541) 498-2531, www.fs.usda.gov/recarea/umpqua/recreation/recarea/?recid=63040.

Finding the springs: From Roseburg head east on OR 138 for 60 miles. Turn north onto FR 34 for 1.4 miles to the Toketee Ranger Station. (Stop at the ranger station to pick up campground information and to purchase a Northwest Forest Pass if you don't already have one.) Continue north past the ranger station on FR 34 for 2.2 miles, passing over 2 concrete bridges that span the Clearwater River. Turn northwest onto FR 3401 and drive 2 miles until you see the Umpqua Hot Springs parking area on the left side of the road. (These last 2 miles may be blocked by a gate in the winter, which will force you to hike.) Park your car and put your Northwest Forest Pass on the dashboard. Take the hiking trail north of the parking area across the footbridge spanning the North Umpqua River and hike 0.1 mile to a trail junction. You're now on the North Umpqua River Trail, a popular hike that parallels the river for 80 miles. Turn right at the junction and hike next to the river, heading upstream, for 0.2 mile to the hot springs. The last few hundred yards of the trail are fairly steep.

THE HOT SPRINGS

Few hot springs in the Pacific Northwest are as scenic as Umpqua. Over thousands of years the mineral water on the banks of the North Umpqua River has deposited a travertine mound more than 100 feet tall. The hot springs emerge from the top of this terrace and flow into four soaking pools.

The main pool, enclosed by a three-sided log shelter, is approximately 5 feet wide by 8 feet long and 2 to 3 feet deep. It can easily hold four or five people. This is the most popular pool, both because of the comfortable soaking temperature and the view of the river valley. The hot springs source is located in a fissure in the travertine about 10 yards above the main pool. The hot water is piped from the rock fissure to the pool through a rubber hose. If you're squeamish about bathing in the shallow soaking pool in hot water that has been used by other bathers, you can use one of the 10-gallon buckets next to the shelter to bail out the water, then let the pool refill with fresh hot water from the hose.

Above the sheltered pool is a smaller (and hotter) pool that holds two to three people. This pool is often empty because the water sometimes approaches 115 degrees F, a bit too toasty for most visitors. An even smaller pool takes the overflow from the small pool, and it is somewhat cooler. There's a fourth pool clinging to the side of the terrace about 20 yards downhill toward the river that collects the runoff from the other three pools. Be careful when walking around the pools on the terrace, especially during rainy weather when the travertine is quite slick. It's a long drop to the river valley below.

History

According to local county history, a settler named Perry Wright homesteaded near the hot springs. Wright's wife, Jessie, recalled that Native Americans used to bathe in the natural basin of hot water. The Wrights would also bathe in the springs in summer, when they were pasturing their cattle in a nearby meadow.

In the early 1900s a Forest Service employee named Carlos Neal was stationed at a fire lookout near the hot springs. Neal would often soak in the shallow natural soaking pool but was dissatisfied with its depth. He decided to enlarge the pool and packed

Steam rising from Umpqua Hot Springs WAYNE ESTES

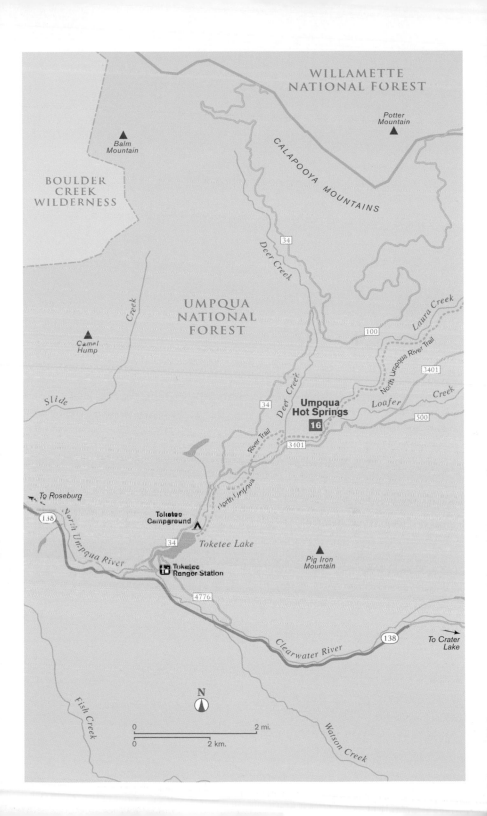

WILLAMETTE
NATIONAL FOREST

Potter
Mountain

Balm
Mountain

CALAPOOYA MOUNTAINS

BOULDER
CREEK
WILDERNESS

Deer Creek

34

UMPQUA
NATIONAL
FOREST

Laura Creek

100

North Umpqua River Trail

Camel
Hump

3401

Creek

Slide

Deer Creek

34

Umpqua
Hot Springs

16

Loafer

500

River Trail

3401

To Roseburg

North Umpqua

138

North Umpqua River

Toketee
Campground

34   Toketee Lake

Toketee
Ranger Station

Pig Iron
Mountain

4776

Clearwater River

138

To Crater
Lake

N

Fish Creek

Watson Creek

0   2 mi.
0   2 km.

Soaking pools cascade down the terrace at Umpqua Hot Springs. WAYNE ESTES

a hammer and chisel with him during his visits to the travertine terrace. Neal chiseled a 3-foot-deep soaking basin into the soft rock during the several summers that he was stationed in the area. The Forest Service later built a three-sided log shelter over the hot springs. Over the years the bark has worn away from the logs that make up the walls of the shelter, replaced with hundreds of initials carved into the wood by visitors to the hot springs.

It's doubtful that Forest Service employees today would be allowed to "improve" natural hot springs in the same way that Neal did in the early 1900s. Fortunately the mineral-water deposits in the subsequent decades have softened the rough chisel marks on the edges of the pool, and bathers usually don't realize that the soaking basin is anything but natural.

In the 1990s three other smaller soaking pools were built, two of them located behind the sheltered pool. A third pool was built in the runoff channel about 30 yards down the travertine terrace's steep slope.

17. JACKSON WELLSPRINGS

General description: A historic mineral-water swimming pool and campground that has been rebuilt into a holistic health retreat.

Location: Southwest Oregon, 2 miles north of Ashland.

Development: The hot springs have been used commercially since the late 1800s. Jackson Wellsprings considers itself to be an "education and healing center and eco-resort," with daily classes and weekend workshops that encourage sustainability and human development.

Best time to visit: The Jackson Wellsprings campground and swimming pool are open year-round.

Restrictions: A fee is charged to use the swimming pool. The facility is clothing optional after dark. The private hot tubs are clothing optional. The pool is closed from midnight Sun to 6 p.m. Mon. Mon evenings the pool is reserved for women only.

Access: Jackson Wellsprings is adjacent to OR 99, so any vehicle can make the trip.

Water temperature: The hot pool measuring 14 x 26 feet is maintained at 100 degrees F. during the day and increased to 104 degrees F. at night. Temperatures in the 45 x 90 foot swimming pool range between 72 degrees F. in February and 86 degrees F. in August.

Nearby attractions: Ashland offers a wide variety of cultural activities that include the Oregon Shakespeare Festival, which runs from Feb through Oct, several music festivals, and a host of art galleries and museums. Less than 10 miles north of Jackson Wellsprings lies the gold-rush town of Jacksonville, which has more than 80 buildings on the National Register of Historic Places. Jacksonville is home to the Britt Festival, a summer-long outdoor music and performing arts gathering that features nationally known artists.

Services: Jackson Wellsprings has 20 RV sites with full hookups, as well as teepees for overnight guests. Restrooms, showers, and laundry facilities are also available.

Camping: WellSprings offers tent and car camping. Dog-free and dog-friendly camping areas are provided.

Maps: Oregon State Highway Map; *DeLorme: Oregon Atlas & Gazetteer,* page 68, C4.

GPS coordinates: N42.2214' / W122.7445'

Contact info: Jackson Wellsprings, 2253 Hwy. 99 N., Ashland, OR 97520; (541) 482-3776; www.jacksonwell springs.com.

Finding the springs: From downtown Ashland drive 2 miles north on OR 99 to a stoplight and junction with I-5. Continue north on OR 99 another 200 yards. Look for a sign marking the turn to Jackson Wellsprings on the left side of the road. Turn into the lane and drive about 50 yards to the Wellsprings pool building.

THE HOT SPRINGS

Jackson Wellsprings is still known to many locals by its old name of Jackson Hot Springs. The name was officially changed in 1995 by its two new owners: Gerry Lehrburger, a physician who specializes in "integrative and transitional medicines," and Bruce Blackwell, an acupuncturist and Chinese herbalist. Lehrburger and Blackwell have been slowly transforming the resort into a nonprofit health and research center.

History

Wellsprings is one of the few hot springs resorts remaining from the heyday of Ashland's mineral springs era. In 1886 a visitor to Ashland described the importance of mineral water to the area:

> The region is particularly rich in soda and sulfur springs, both cold and thermal, and is fast becoming a resort for invalids, tourists, and those in search of the healing waters of nature. The opening of through railroad communications will, in the near future, bring thousands from abroad to the fountains of life.

Dome tents provide an upscale camping experience at Jackson Wellsprings. LINDSAY WOLTER

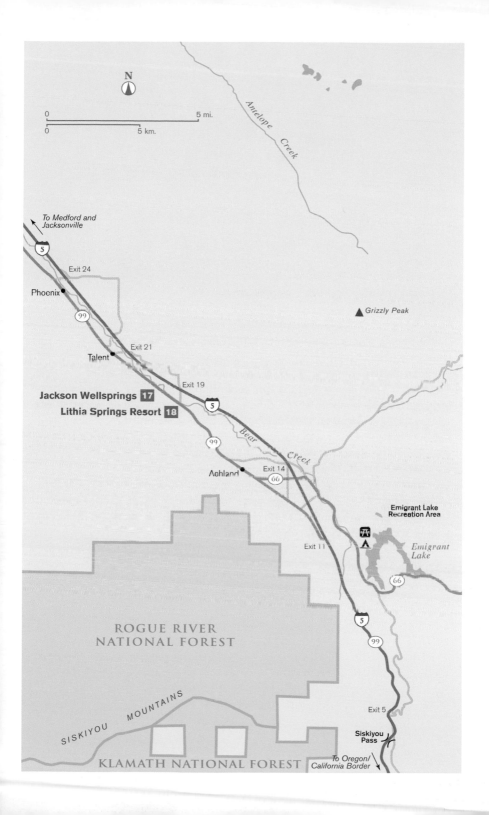

The outdoor pool at Jackson Wellsprings LINDSAY WOLTER

Several hot springs and mineral springs resorts developed in and around Ashland, including Wagner's Soda Springs, Buckhorn Springs, Helman Baths, White Sulphur Springs, and Kingsbury Springs.

John Barrett claimed Jackson Hot Springs and the surrounding land in 1857. Thirty years later Barrett sold the land to G. F. Billings. Billings's granddaughter, Eugenia Jackson, eventually inherited the hot springs and surrounding ranchland. She sold the springs to Jacob and Gertrude Ottinger, on the condition that the name "Jackson Hot Springs" remain with the property. The Ottingers built a public swimming pool, an outdoor dance pavilion, and small rental cabins.

Unfortunately all these buildings burned to the ground in 1933.

The Ottingers rebuilt the pool building and cabins and continued to manage the hot springs until 1960, when the property was sold to William Wallace and Ruth Smith, who managed the facility into the 1980s. In 1995 the resort was sold to the Jackson Wellsprings LLC, the current owners.

18. LITHIA SPRINGS RESORT

See map on page 85.

General description: A charming country resort with naturally heated mineral-water whirlpool tubs in the guest cottages. The resort is a great getaway for a romantic weekend, which could include attending the Oregon Shakespeare Festival in nearby Ashland.

Location: Southwest Oregon, 2 miles north of Ashland.

Development: Lithia Springs Resort was first established in 1991.

Best time to visit: The resort is open year-round. During Ashland's popular Oregon Shakespeare Festival (Feb–Oct), room reservations should be made as far ahead of time as possible.

Restrictions: The spring-fed soaking tubs in the cottages are reserved for overnight guests of the resort.

Access: Any vehicle can make the trip—the resort is right off OR 99.

Water temperature: The well behind the resort provides 96-degree-F mineral water to the resort. The water temperature is boosted another 20 degrees F prior to being piped to the soaking tubs in the guest cottages.

Nearby attractions: Ashland offers a wide variety of cultural activities that include the Oregon Shakespeare Festival, which runs from Feb through Oct, several music festivals, and a host of art galleries and museums. A few miles north of the resort lies the gold-rush town of Jacksonville, which has more than 80 buildings on the National Register of Historic Places. Jacksonville is home to the Britt Festival, a summer-long outdoor music and performing arts gathering that features nationally known artists.

Services: The resort at Lithia Springs opened in 1991. Guest accommodations include 38 bungalows, suites, and studio apartments. Each room contains a private soaking tub that can be filled with natural mineral water. The resort also features an outdoor saline swimming pool during the warmer months, as well as an outdoor Jacuzzi and a sitting area by a koi pond. Facial massages and skin treatments are available by appointment. A full breakfast is included with the room.

Camping: No camping is available at the resort, but camping can be found at Jackson Wellsprings, about 0.5 mile north of Lithia Springs Resort.

Map: DeLorme: Oregon Atlas & Gazetteer, page 68, C4.

GPS coordinates: N42.2197' / W122.7425'

Contact info: Lithia Springs Resort, 2165 W. Jackson Rd., Ashland, OR 97520; (541) 482-7128; www.lithia springsresort.com.

Finding the springs: From downtown Ashland drive 2 miles north on OR 99. Turn left onto West Jackson Road and drive 0.1 mile to the resort.

THE HOT SPRINGS

While nearby Jackson Wellsprings caters to a more relaxed holistic crowd, the Lithia Springs Resort targets a better-heeled clientele.

Most of the suites and cottages contain whirlpool baths that can be filled with slightly sulfurous mineral water. The thermal water leaves a very soft and silky feel to the skin. One couple wrote in the guest book that the mineral-water whirlpool enhanced their "carnal knowledge" during their stay. Another guest, a photographer from Finland, praised the curative effects of the water. The photographer had suffered a rash on her nose since childhood, but after four days of soaking in the mineral water at the resort, the rash disappeared. According to the proprietor, the woman carried 2 gallons of the mineral water with her on the plane home to Finland.

Lithia Springs Resort offers a full breakfast with freshly baked pastries and one or two special main dishes, as well as an assortment of juices, teas, and cereals. All the herbs and many of the vegetables included in the breakfast entrees are grown in the garden behind the inn.

Although the resort doesn't offer lunch or dinner service, nearby Ashland features more than a dozen top restaurants.

History

In the early twentieth century, the Ashland area was famous for its many mineral and hot springs resorts. A 1935 county directory touted the curative effects of these waters:

Ashland is famous for its mineral waters. Foremost among these is the Lithia Water—a mineral water that has gained for Ashland the name of "Lithia City."

A private soaking tub in a spacious guest room at Lithia Springs Resort ALEXANDER NEUMAN

The tranquil pond and garden at Lithia Springs Resort ALEXANDER NEUMAN

Famed for its curative powers, it is also a delightful beverage and made available in hotels and fountains, the chilled Lithia Water becomes one of the big attractions of the "Lithia City." People suffering from rheumatism, stomach, liver, intestinal and other ailments come from all parts of America for treatment and baths.

Ashland still offers "lithia water" at the public fountain near the hundred-acre Lithia Park. Like all of Ashland's downtown attractions, the park is easily accessible from Lithia Springs Resort via a 2-mile-long bicycle and walking path. Ashland is well known for the Oregon Shakespeare Festival, which attracts more than 100,000 playgoers during its nine-month season.

CENTRAL OREGON

19. PAULINA LAKE HOT SPRINGS AND EAST LAKE HOT SPRINGS

General description: Hot springs seeping into the gravel beaches of 2 lakes located in the Newberry National Volcanic Monument. Bathers dig down into the gravel to create soaking pools that fill with a mixture of hot springs water and colder lake water.

Location: Central Oregon, on the shores of Paulina Lake and East Lake in the Newberry National Volcanic Monument, 20 miles southeast of Bend.

Development: None, other than logs and rocks that are used to create the walls of soaking pools on the beaches.

Best time to visit: Late spring and summer. High water levels in the lakes in winter and early spring can dilute the beachside hot springs, rendering them too cold for soaking. Snow may close the roads to Newberry Crater and the hot springs during much of the winter.

Restrictions: Overnight camping is not permitted on the beaches near the warm springs. Clothing is optional, but fishing boats are often within eyesight just offshore from the hot springs, so use discretion. A National Forest Recreation Pass is required to visit the Newberry National Volcanic Monument.

Access: Any vehicle can make the trip to the parking areas by Paulina Lake and East Lake. The hikes along the shore to reach the hot springs beaches are easy.

Water temperature: The hot springs along Paulina beach are just below the gravel and sand, with temperatures from 95 to 115 degrees F. The temperature of the hot springs on the beach at East Lake is similar to the Paulina Lake hot springs. The actual temperature in any soaking pool depends on the season. The temperature of the hot springs vents deep within the lakes may exceed 170 degrees F, but the lake water cools off this scalding water to a comfortable temperature for soaking by the time it reaches the gravel of the beaches.

Nearby attractions: The Newberry National Volcanic Monument, which contains both Paulina Lake Hot Springs and East Lake Hot Springs, has a variety of interesting geology, as well as excellent fishing, camping, and hiking opportunities. The lakes are home to trophy-size brown and rainbow trout and both kokanee and Atlantic salmon. The monument encompasses a 17-square-mile depression called a caldera, which formed when the summit of the volcano collapsed around 500,000 years ago. A shallow magma body of molten rock may lie only a few kilometers beneath the volcanic caldera, and the hot springs and interest in geothermal development just outside the monument boundary shows that the volcano is far from slumbering. Other volcanic remnants within the monument include over 400 cinder cones and a number of ancient basalt flows.

Services: East Lake Resort, on the shores of East Lake, offers RV and tent sites, cabins, and boat rentals.

Camping: Several campgrounds are available around Paulina Lake and East Lake, including Little Crater Campground, Paulina Lake Campground, East Lake Campground, and Cinder Hill Campground. No reservations are taken for the campgrounds, so arrive early to pitch your tent if you want to ensure you get a spot.

Maps: Oregon State Highway Map; *DeLorme: Oregon Atlas & Gazetteer,* page 50, C4.

GPS coordinates: N43.7173' / W121.2090'

Contact info: Deschutes National Forest, 63095 Deschutes Market Rd., Bend, OR 97701; (541) 383-5300; www.fs.usda.gov/recarea/deschutes/recarea/?recid=71997&actid=42.

Finding the springs: From Bend drive 23 miles south on US 97. Turn east onto CR 21 (also called Paulina East Lake Road) and drive 17 miles to the Newberry Crater. To reach the Paulina Lake Hot Springs, continue driving to Paulina Lake to the Little Crater boat ramp. Park your car and hike 2.5 miles along the Paulina Lakeshore Loop Trail. Look for signs of digging and pool construction in the beach and you'll know you've arrived. To find the East Lake Hot Springs, continue driving past Paulina Lake to the parking ramp at Hot Springs boat ramp on East Lake. Park your vehicle and hike along the beach to the west for about a hundred yards (look for an outcropping of tan rocks just behind the beach). You may smell the characteristic "rotten egg" smell of hydrogen sulfide in the beach, which indicates you've arrived at the hot springs area.

A hot soaking pool dug in the sand on the shores of Paulina Lake WAYNE ESTES

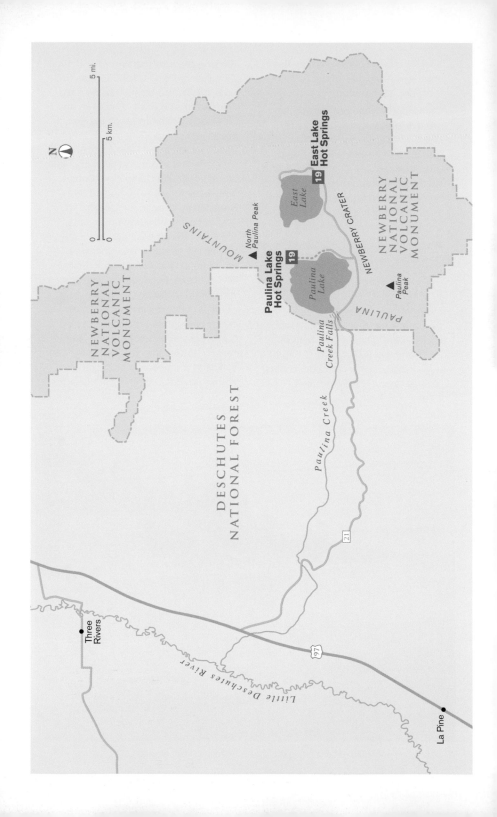

5 mi.

5 km.

N

Newberry National Volcanic Monument

East Lake

19 **East Lake Hot Springs**

North Paulina Peak

Paulina Lake Hot Springs **19**

Paulina Lake

Newberry National Volcanic Monument

Newberry Crater

Paulina Peak

Paulina Mountains

Paulina Creek Falls

Paulina Creek

Deschutes National Forest

21

Three Rivers

Little Deschutes River

97

La Pine

Scenic view from a pool dug in the beach sand at Paulina Lake WAYNE ESTES

THE HOT SPRINGS

The hot springs of Paulina Lake and East Lake require a bit of work before you can enjoy a soothing soak. The hot water emerges from deep within the lakes, then cools and floats near the lake surface, penetrating the sand and gravel along the lakeshores. In two specific areas of Paulina Lake and East Lake, these hot-water beaches are easily accessible, and visitors with a little time and a shovel can dig down to create their own soaking pool.

Often you'll find a pool already built, surrounded by rocks or logs to contain the hot water. But you may need to deepen the hole to reach hotter water or enlarge the soaking areas. The water levels in the lake also can impact your soaking pleasure—high water can totally wash out the soaking areas, or at least cool the hot water down so much that it's too chilly for a soak.

History

The East Lake Health Resort used the hot springs at East Lake for almost thirty years, from 1913 to 1941. Access to the resort in the early days climbed a steep road from the east up the side of the Newberry Caldera. The resort originally had just four cabins and a small bathhouse. A larger health resort was planned but never built. No traces of the old health resort remain.

20. SUMMER LAKE HOT SPRINGS

General description: A hot-water plunge enclosed in an old metal-clad barn that's changed little since 1928. A variety of overnight accommodations and an isolated location add to the resort's charm. With almost no development or other artificial lights in 50 miles in any direction, Summer Lake Hot Springs is a wonderful base for stargazing.

Location: South-central Oregon, 51 miles northwest of Lakeview.

Development: A rustic bathhouse built in the late 1920s encloses the hot springs pool. The 143-acre property contains RV and camping spots, cabins, and a guesthouse.

Best time to visit: Feb and Mar are peak months for viewing migrating snow geese on nearby Summer Lake. In midsummer hang-gliding aficionados launch from nearby Winter Ridge, some 4,000 feet higher than the surrounding valley, and land on the hot springs property. Stargazers who come to view meteor showers streaking across the night skies prefer autumn. In recent years Summer Lake Hot Springs has become an overnight stop for revelers on their way to and from the Burning Man celebration in the Nevada desert, which is held the last week of Aug and first week of Sept. Up to a thousand visitors a day have descended on the resort during the Burning Man pilgrimage. The resort is also busy in the fall during deer-hunting season.

Restrictions: A fee is charged for day use of the hot springs. Soaking is free to overnight guests. Swimsuits are required in the pools before 9 p.m. After 9 p.m. it's clothing optional.

Guests have 24-hour access to the pools.

Access: Any vehicle can make the trip to Summer Lake Hot Springs, which is located just off OR 31. The resort is open year-round.

Water temperature: Three separate hot springs, averaging 113 degrees F, are piped to the bathhouse, private hot tubs, and RV park hookups. A nearby cold spring is piped to the bathhouse for cold showers. The swimming pool in the bathhouse is maintained at a comfortable 102 degrees F in winter and 98 to 100 degrees F in summer. The 3 outdoor pools range from 104 to 107 degrees F.

Nearby attractions: The nearby town of Paisley hosts a Mosquito Festival every summer (bring your own bug repellent). Summer Lake Wildlife Area, located 22 miles north of Summer Lake Hot Springs on OR 31, is an ideal place to spend a spring day watching migrating waterfowl. Hart Mountain National Antelope Refuge lies about 50 miles east of Summer Lake. The Chewaucan River, which flows through Paisley, is well known to area anglers, as are the nearby Sycan, Sprague, and Ana Rivers. Ancient rock drawings can be seen at Picture Rock Pass, about 30 miles north of the hot springs.

Services: No services are available other than lodging, soaking, and enjoying nature. Bring your own food and beverages if you're spending the night at the campground or in one of the cabins or lodge facilities. You can pick up provisions in Paisley, 6 miles south of the hot springs. Only natural mineral water is available at the resort (which may not be to your liking), so bring your own drinking water.

Camping: Summer Lake Hot Springs has a variety of lodging options, including tent sites, over a dozen RV sites with full hookups, 4 cabins, and a 2-bedroom guesthouse that sleeps 6. The cabin walls are made of a "pumicecrete, " a lightweight volcanic rock found around Summer Lake from the eruption of Crater Lake over 7,000 years ago. Many of the guesthouses are also heated with geothermal energy from the hot springs. This naturally heated water runs through pipes in the floors, keeping them warm and toasty. Camping is also available at the Summer Lake Wildlife Area, 22 miles north of the hot springs on OR 31.

Maps: Oregon State Highway Map; *DeLorme: Oregon Atlas & Gazetteer,* page 84, B3.

GPS coordinates: N42.7258' / W120.6472'

Contact info: Summer Lake Hot Springs, 41777 Hwy. 31, Paisley, OR 97636; (541) 943-3931; www.summer lakehotsprings.com; duane@summer lakehotsprings.com.

Finding the springs: From Lakeview drive 22 miles north on US 395 to the junction with OR 31. Turn left onto OR 31 and drive 23 miles northwest to the small town of Paisley. Drive beyond Paisley for 6 miles on OR 31 to milepost 92 and turn at the Summer Lake Hot Springs sign into the resort driveway.

An old wooden sign points the way to Summer Lake Hot Springs.

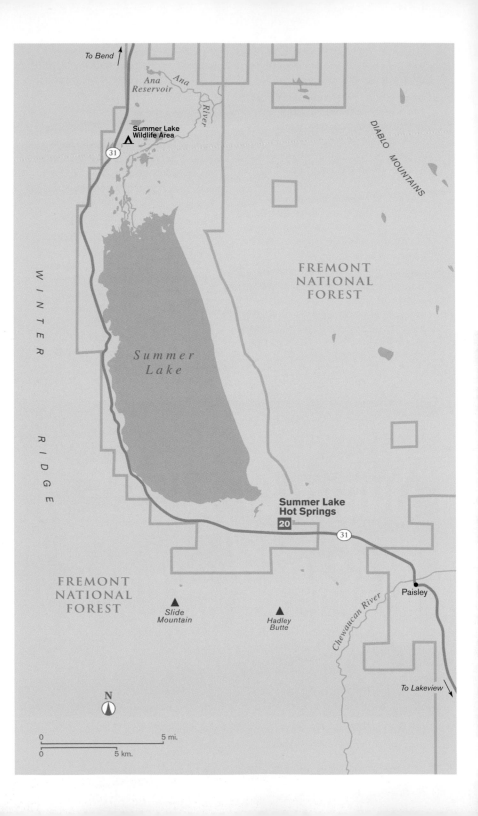

THE HOT SPRINGS

One visitor in the 1970s likened the barnlike building that encloses the hot-water pool to "an overgrown chicken coop." The cavernous structure is dimly lit through windows on the west end of the building, and shafts of sunlight that penetrate cracks in the corrugated-steel roof dance on the surface of the warm-water pool.

The 15-by-30-foot pool ranges from 3.5 to 5 feet in depth. About 20 gallons per minute of 113-degree-F hot water is piped into one end of the pool, where it mixes with colder well water to create a comfortable soaking temperature. The pool is flanked by white wooden dressing rooms (men on one side, women on the other). Three outdoor, rock-lined soaking pools are snuggled against the north end of the bathhouse.

History

Explorer John C. Frémont ventured into the valley surrounding the hot springs in 1843 during a reconnaissance expedition to establish a route for the Oregon Trail. Noting the balmy climate of the area, Frémont coined the name Summer Lake for

The original bathhouse at Summer Lake Hot Springs, built in 1928

Outdoor soaking pools at Summer Lake Hot Springs

the shallow body of water north of the hot springs. Many settlers soon followed in Frémont's footsteps, most continuing on west but some staying in the sunny valley. By 1873 the town of Paisley, 6 miles south of the hot springs, had its own post office.

Old legends tell of arthritic neighbors being carried by wagon to the undeveloped hot springs, where they soaked in the surrounding hot, black mud to alleviate their aches and pains.

The first individual to exploit the economic potential in the hot water was Jonas Woodward, who owned the property in the early 1900s. Woodward carved bathtubs out of large logs that he then filled with hot water for private baths. A wood-lined swimming pool and wooden bathhouse eventually replaced the log tubs. In the early

The large indoor pool at Summer Lake Hot Springs

1920s George Minton purchased the hot springs from the Woodward family. In 1928 Minton's son Claude replaced the original wooden bathhouse with a concrete pool and the corrugated-steel bathhouse. This barn-shaped structure, complete with a pitched roof and gables that let the steam escape from the pool, has welcomed bathers for nearly nine decades.

In the mid-1930s the Mintons sold the well-known little resort. The hot springs had several owners until the late 1950s, when Jeff McDaniel purchased the property. The springs grew in popularity during the forty years that the McDaniel family managed the hot springs. McDaniel stated in a 1970s news article that the old hot-water plunge had so many summer visitors that "there's sometimes no place to wiggle a toe." The McDaniel family operated the hot springs resort until 1997, when they sold the property to Duane Graham.

21. HUNTER'S HOT SPRINGS RESORT

General description: A historic resort with an outdoor thermal soaking pool on the outskirts of Lakeview. The resort is widely known for its man-made geyser ("Old Perpetual"), which has erupted every 60 seconds for over 90 years.

Location: South-central Oregon, 2 miles north of Lakeview on US 395.

Development: The resort has been developed since the early 1900s.

Best time to visit: Hunter's Hot Springs is open year-round. Crisp autumn days are a good time to visit to see flocks of migrating birds that stop at the ponds near the resort.

Restrictions: Swimsuits are required in the warm-water public pool. Resort guests swim for free; others must pay a day-use fee. Private hot tubs are available outside of some of the ground-floor hotel rooms—these are clothing optional.

Access: Any vehicle can travel the paved highway to the resort. The outdoor pool is open daily from 10 a.m. to 10 p.m. but is closed Tues evening after 8 p.m. for cleaning.

Water temperature: Hot mineral water flows into an open-air swimming pool at a scalding 185 degrees F, but 2 hoses supply cold water to the pool, which keeps it at a comfortable 95 to 100 degrees F.

Nearby attractions: Backpackers and mountain bikers can take advantage of the nearby Crane Mountain National Recreational Trail, a scenic 31-mile-long ramble off OR 140 that extends south to the Oregon-California border. Five miles east of Lakeview is Black Cap Overlook, which rises 2,000 feet above the city.

Black Cap is a favorite hang-gliding launching area. Warner Canyon Ski Area is about a 10-minute drive east of the hot springs on OR 140.

Services: The resort features a lounge, motel, outdoor pool, and bird-watching. Gas, groceries, and more dining and lodging options are available in Lakeview, 2 miles south of the resort. Current owner Mike Watson purchased the property in 2018 and has been slowly remodeling the sleepy resort. Future plans include a bar and a restaurant, refurbished rooms, and private soaking tubs. Much of this remodeling is expected to be finished by the summer of 2021. Hunter's Hot Springs Resort makes a nice base of operations if you have a long hot springs weekend planned. Consider staying at the resort and conducting day trips to nearby Summer Lake Hot Springs, Crump Geyser, Fisher Hot Springs, and Antelope Hot Springs.

Camping: An RV park and campground is adjacent to the resort.

Maps: Oregon State Highway Map; *DeLorme: Oregon Atlas & Gazetteer,* page 84, E4.

GPS coordinates: N42.2192' / W120.3648'

Contact info: Hunter's Hot Springs Resort, 18088 N. Rte. 395, Lakeview, OR 97630; (541) 947-4242; (no website as of Aug 2020).

Finding the springs: From downtown Lakeview drive north 2 miles on US 395. You'll see the resort just off the highway to the left. Turn at the Geyser/Hunter's Hot Springs sign and park by the main lodge. (Some locals still refer to the resort as Geyser Hot Springs Resort; most call it Hunter's Hot Springs.)

THE HOT SPRINGS

Just north of the main lodge sits a shallow pond containing the main attraction at Hunter's Hot Springs. "Old Perpetual" is a man-made geyser that has erupted regularly for over ninety years. Since the early 2000s the eruptions have been a bit less dependable, as an ongoing drought, irrigation pressures on the water table, and geothermal energy development nearby appear to be taking their toll on the geyser's regularity. The surrounding duck ponds, which are fed by the geyser, sometimes dry up in the late summer due to Oregon's ongoing drought.

A neighbor tells an interesting story about Goldfish Lake, the warm-water pool surrounding the geyser. Years ago an owner of the resort stocked the pool with exotic Japanese koi, which grew quite large in the thermal waters. One autumn a migrating flock of pelicans landed on the pool and proceeded to devour the entire population of expensive fish, much to the chagrin of the resort owner.

Fortunately for resort guests there's another source of thermal water besides the geyser in Goldfish Lake. A separate well supplies 185-degree-F water to a swimming pool behind the main lodge. The pool is 15 feet by 30 feet and varies from 3 to 5 feet

The warm-water pool at Hunter's Hot Springs Resort

deep. Hot water enters through a hose in the pool's deep end. Two other hoses supply cold water to keep the pool at a comfortable temperature.

History

In 1832 a trapper for the Hudson's Bay Company discovered the hot springs. In his journal he described the steaming water: "There is a hot spring in the valley a little to the side of the trail. Some of the young men went to it, and found the water so hot that the finger can barely endure in it a moment. There are a number of human skulls and other bones in it but how they came there is no knowing."

There is little recorded history about the hot springs for the next eighty-seven years, until a wealthy land developer from Minneapolis named Harry Hunter visited the area in 1919. Hunter had traveled through Boise, Idaho, on his trip west and had seen hot springs used to heat buildings in that city.

When Hunter saw the hot springs north of Lakeview, he realized the commercial potential of developing the property. Hunter purchased the hot springs in 1923 and began plans to build a sanatorium, a public swimming pool, and a golf course.

The flow of the artesian hot springs wasn't sufficient to supply enough thermal water for Hunter's visions, so he hired a well-drilling company to try to find more hot water. Three wells were drilled in 1923, and to Hunter's surprise, they not only found hot water, but all three wells erupted as man-made geysers. Eventually two of the wells died down, but the third well, which was named Old Perpetual, has continued to erupt to this day.

Pool sign at Hunter's Hot Springs Resort

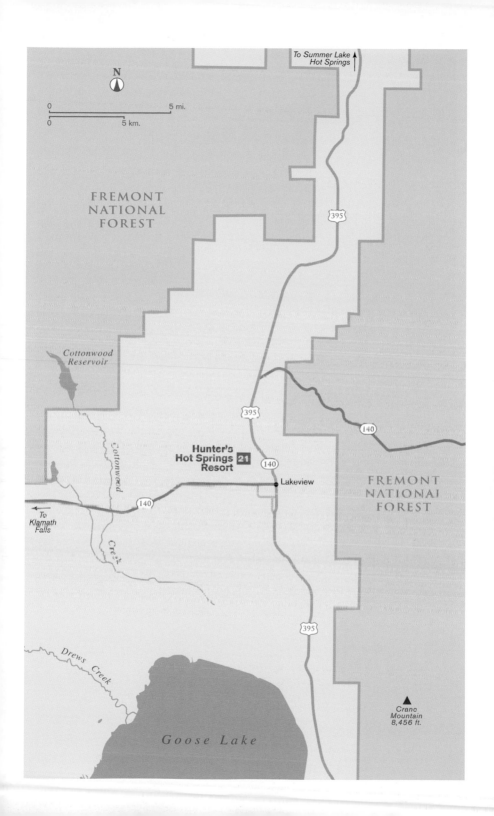

N

0 ——————— 5 mi.
0 ——————— 5 km.

To Summer Lake
Hot Springs

FREMONT
NATIONAL
FOREST

395

Cottonwood
Reservoir

395

Cottonwood

Hunter's
Hot Springs 21
Resort

140

140

Lakeview

To
Klamath
Falls

140

Creek

FREMONT
NATIONAL
FOREST

Drews Creek

395

Crane
Mountain
8,456 ft.

Goose Lake

Old Perpetual Geyser at Hunter's Hot Springs Resort JAMES GULLICKSON

Hunter was elated with the amount of hot water he found, so he set about to raise capital to build his resort. The experienced developer formed the Hunter's Chlorine Hot Springs Club as a business entity, sold common stock, and raised $100,000 for construction. Hunter built a twenty-two-room sanatorium with an indoor pool for patients and an outdoor pool for recreational use by vacationers. Hot water from one of the wells was used to heat the building, but the geyser was left untouched so that visitors could enjoy its constant display.

Unfortunately Hunter died six years after purchasing the property, and the assets of his Hunter's Chlorine Hot Springs Club were soon liquidated.

Over the next seventy years, the resort had a variety of owners. Mike Watson purchased the property in 2018, and it is managed by his son Sam. The Watsons have been slowly updating the old resort, with plans for most updates to be finished by 2021.

22. ANTELOPE (HART MOUNTAIN) HOT SPRINGS

General description: A high-elevation, primitive hot springs located in the center of the Hart Mountain National Antelope Refuge.

Location: South-central Oregon, 68 miles northeast of Lakeview.

Development: Hot water bubbles into a natural rock soaking pool. Rock walls and a cement deck surround the springs. A second open pool about 50 yards west from the enclosed spring also offers a good soaking experience.

Best time to visit: Situated just shy of 6,000 feet in elevation, Antelope Hot Springs is best visited in summer and autumn, as winter snow can linger well into spring. The hot springs are especially popular on weekends, holidays, and during fall hunting season, but midweek visitors will rarely find a crowd.

Restrictions: A hand-painted sign on the bathhouse door used to request that soaks be limited to 20 minutes when others are waiting, and to use discretion when bathing in the nude. In practice it seems rare that anyone actually cares if others are soaking in the buff. It's probably best to holler before entering the bathhouse to see if it's occupied.

Access: Most vehicles can drive the 64 miles from Lakeview to the refuge headquarters, although the last 3 miles switchback more than 2,000 feet up a narrow road to the refuge plateau. The 4-mile road on the plateau from refuge headquarters to Antelope Hot Springs is bumpy and poorly maintained, and RV and trailer use is not recommended. Winter snows can sporadically close the road to the hot springs from Oct through May, although the main refuge road remains open year-round.

Water temperature: The temperature varies between 99 and 102 degrees F in the enclosed soaking pool. The open meadow pool about 50 yards west of the enclosed pool is a few degrees warmer (around 104 to 106 degrees F).

Nearby attractions: The 275,000-acre Hart Mountain National Antelope Refuge surrounding the hot springs provides grass for pronghorn antelope herds in spring and summer. The current antelope population, numbering around 1,200 animals, migrates south and east to lower elevations in the winter months. Since its creation the refuge has broadened its mission to include conservation of other wildlife and native plant species.

The refuge sits atop a massive fault block ridge that rises abruptly more than 3,600 feet from the floor of the Warner Valley. Bighorn sheep inhabit many of the rocky crags along the west side of the refuge. The ridge slopes gently to the east, past the refuge headquarters and farther, more than 40 miles to OR 205. Much of the plateau consists of sagebrush flats that stretch to the horizon.

Animals common to the refuge include coyote, deer, and more than 200 species of birds. It's not uncommon to see a dozen or more pronghorn antelope near the road that bisects the eastern portion of the refuge. One of the swiftest animals on earth, pronghorn can easily reach a speed of 60 miles an hour, much faster than you could possibly drive on the rough washboard road on the plateau. If you are traveling east from the refuge headquarters, leave your speeding antelope fantasies behind and allow a couple of hours to drive the bumpy 40 miles to OR 205.

In addition to wildlife-watchers, the refuge is popular with backpackers, anglers, and hunters. Collecting rock specimens is also a popular activity. The refuge allows rock hounds to collect up to 7 pounds of rocks per person per day.

Services: None at the springs, except for a nearby pit toilet. Bring plenty of water and food if you're visiting for the day, and bring all needed camping gear for overnight stays. Gasoline and groceries are available in Plush (28 miles west) and Frenchglen (49 miles east). The Frenchglen Hotel is an excellent historical overnight accommodation. Lakeview (68 miles southwest) is the nearest full-service community with overnight lodging.

Camping: The hot soaking pools are surrounded by Hot Springs Campground. No camping permits are required here, but it's still wise to stop at the refuge headquarters to check on current conditions before driving the final 4 miles to the hot springs. The campground has 30 primitive camping spots but no drinking-water source (fill up your water bottles at a spigot near the refuge headquarters building). There's a 14-day limit on overnight camping. Camping is also allowed elsewhere on the refuge (especially during special refuge hunting seasons), but a camping

permit must be purchased at the refuge headquarters. Backpackers on overnight trips must also obtain a camping permit.

Maps: Oregon State Highway Map; Hart Mountain National Antelope Refuge Recreational Use Map; *DeLorme: Oregon Atlas & Gazetteer,* page 85, C7.

GPS coordinates: N42.5010' / W119.6918'

Contact info: Hart Mountain National Antelope Refuge, US Fish and Wildlife Service, National Wildlife Refuge System, 20995 Rabbit Hill Rd., PO Box 111, Lakeview, OR 97630; (541) 947-3315; www.fws.gov/sheldonhartmtn/hart/management.html.

Finding the springs: Most visitors to Hart Mountain National Antelope Refuge drive northeast from Lakeview. From Lakeview head north 40 miles on OR 395 to the town of Plush. Drive 1 mile north of Plush, then turn east at the sign to Hart Mountain. Follow the signs along Hart Mountain Road about 20 miles to the base of the Hart Mountain escarpment. The road then gains more than 2,000 feet as it switchbacks over the next 3 miles up the side of Hart Mountain to the high-altitude plateau. The refuge headquarters will appear soon after you emerge on the plateau. Stop at the headquarters for the latest information on the hot springs and campground conditions, then proceed to the junction just west of the headquarters. Turn south onto the unimproved road to the Hot Springs Campground and drive about 4 miles to the campground and hot springs.

THE HOT SPRINGS

Situated near the head of Rock Creek in an aspen-dotted meadow, the main enclosed pool at Antelope Hot Springs is a relaxing soak for up to half a dozen people. The soaking pool is about 5 feet deep and 9 feet by 12 feet across. Hot water averaging 100 degrees F bubbles gently from the natural sand-and-rock bottom. A small ladder at one end of the pool provides a way to ease yourself into the warm water. A small sitting bench next to the surrounding walls provides a place for your clothes. The natural stone walls provide privacy and protection from winter winds but also block any view of the surrounding meadows and hills. Fortunately there's no roof over the hot springs, so it's a great place for stargazing on a clear night. A smaller (and hotter) soaking pool is somewhat hidden about 50 yards west of the enclosed pool in a meadow. This meadow pool is about a foot deep and 6 feet around—a nice soaking spot for two to four of your closest friends.

The enclosed soaking pool at Antelope Hot Springs WAYNE ESTES

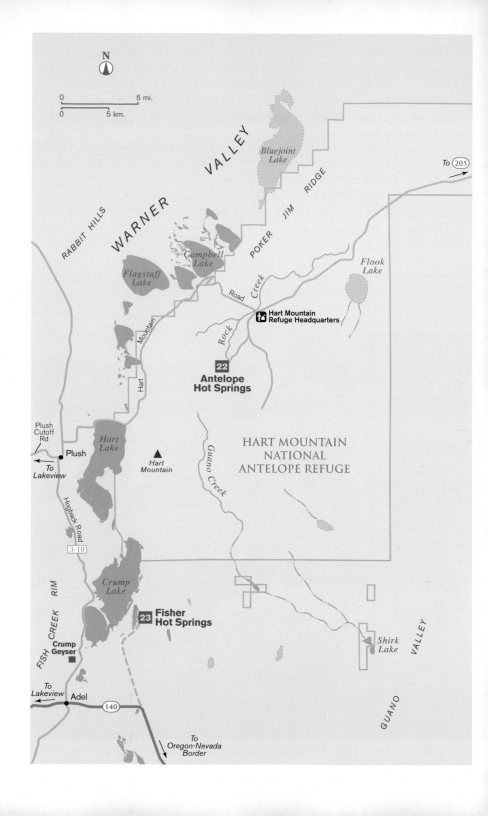

N

0   5 mi.
0   5 km.

VALLEY

Bluejoint
Lake

WARNER

RABBIT HILLS

POKER   JIM   RIDGE

To 205

Campbell
Lake

Flagstaff
Lake

Flook
Lake

Road   Creek

**Hart Mountain
Refuge Headquarters**

Rock   Creek

Mountain

22
**Antelope
Hot Springs**

Hart

Plush
Cutoff
Rd

● Plush

Hart
Lake

▲
Hart
Mountain

Guano   Creek

HART MOUNTAIN
NATIONAL
ANTELOPE REFUGE

←
To
Lakeview

Hogback Road

3-10

FISH CREEK RIM

Crump
Lake

23 **Fisher
Hot Springs**

Shirk
Lake

VALLEY

■
Crump
Geyser

To
Lakeview

Adel ●

140

GUANO

↓
To
Oregon-Nevada
Border

The open-air meadow pool at Antelope Hot Springs WAYNE ESTES

History

Rumor has it that the springs were first "developed" by a rancher who placed a stick of dynamite in the crack where the original artesian hot springs emerged from the ground. The resulting explosion created the current soaking pool.

Soon after the creation of the Hart Mountain National Antelope Refuge in the 1930s, a citizens group was formed to provide volunteer assistance. Called the Order of the Antelope, the group helped remove old barbed-wire fences that were a danger to the free-roaming antelope herds. In the late 1930s the Order of the Antelope also constructed the cement block bathhouse that still surrounds Antelope Hot Springs. Although the Order of the Antelope provided many volunteer services to the refuge, its social events eventually took on a higher priority. By the 1960s the group's annual three day meeting at the refuge was known more for its wild parties than for any volunteer work. The governor of Oregon finally ordered the removal of the buildings that the order had constructed for its annual gatherings, and in 1992 the Order of the Antelope ceased its association with the refuge.

23. FISHER HOT SPRINGS

See map on page 108.

General description: An isolated hot-water soak with gorgeous views of Warner Valley and Hart Mountain.

Location: South-central Oregon, east edge of Warner Valley, 15 miles northwest of Adel.

Development: In past years the soaking facilities at Fisher Hot Springs included an eclectic mix of galvanized stock watering tanks and both fiberglass and cast-iron bathtubs. Currently there is only one soaking option—a large steel tank that can easily hold 2 to 4 bathers. A large tarp is sometimes available to line the tank to reduce leakage.

Best time to visit: The springs are available year-round, although wet weather may prevent access if the dusty roads turn to muck.

Restrictions: None. Nudity is common.

Access: High-clearance vehicles are recommended to travel the final 10 miles of dirt road. Wallows of deep dust are present in the roadbed during the dry season, but these quickly turn to impassable mud during a rainstorm. Don't attempt the drive during rainy weather. The water source for Fisher Hot Springs is actually on BLM land, but the parking lot and actual soaking area are on private land.

Water temperature: The hot springs emerge from a hillside at 154 degrees F. Soaking temperatures in the steel tank will vary, depending on how long you let the hot water cool down before you immerse yourself.

Nearby attractions: The 275,000-acre Hart Mountain National Antelope Refuge lies just to the north of Fisher Springs. The refuge provides grass for pronghorn antelope herds in spring and summer. The current antelope population, numbering around 1,200 animals, migrates south and east to lower elevations in the winter months. Since its creation, the refuge has broadened its mission to include conservation of other wildlife and native plant species.

The refuge sits atop a massive fault block ridge that rises abruptly more than 3,600 feet from the floor of the Warner Valley (Fisher Hot Springs lies on the east edge of this valley). Bighorn sheep inhabit many of the rocky crags along the west side of the refuge. Much of the plateau consists of sagebrush flats that stretch to the horizon.

Animals common to the refuge include coyote, deer, and more than 200 species of birds. It's not uncommon to see a dozen or more pronghorn antelope near the road that bisects the eastern portion of the refuge. One of the swiftest animals on earth, pronghorn can easily reach a speed of 60 miles an hour, much faster than you could possibly drive on the rough washboard road on the plateau.

In addition to wildlife-watchers, the refuge is popular with backpackers, anglers, and hunters. Collecting rock specimens is also a popular activity. The refuge allows rock hounds to collect up to 7 pounds of rocks per person per day.

Services: None at the hot springs. Gas, groceries, and a restaurant are available in Adel, 15 miles to the southwest. Lodging and more restaurant options are available in Lakeview, about 30 miles west of Adel. (Hunter's Hot Springs Resort in Lakeview is a fine base for exploring hot springs in the area.)

Camping: Visitors sometimes camp within 50 yards of the springs (look for a fire pit and signs of previous campers). Camping is allowed (without a permit) on any surrounding BLM land.

Maps: Oregon State Highway Map; *DeLorme: Oregon Atlas & Gazetteer,* page 85, D7.

GPS coordinates: N44.9747' / W118.0448'

Contact info: Bureau of Land Management, Lakeview District Office, 1301 S. "G" St., Lakeview, OR 97630; (541) 947-2177; www.blm.gov/office/lakeview-district-office.

Finding the springs: From Adel drive east on OR 140 for 4.8 miles. Just after you cross a bridge, OR 140 continues to the south, but you need to turn north off OR 140 onto a dirt road that proceeds north along a ridge on the east side of Warner Valley. Head north on this dirt road for 7.5 miles to a junction and take the left fork. Drive another 2 miles past an old ranch. In about another 0.5 mile, you'll see a large rusted tank and a lush creek that forms the runoff from Fisher Hot Springs. Park your car in the open parking area near the tank.

THE HOT SPRINGS

Soaking at Fisher Hot Springs provides you with one of Oregon's prettiest views— the open Warner Valley to the south and west, and the massive Hart Mountain rising from the valley to the north. Soaking here requires a bit of work—the large steel

Fisher Hot Springs with the Warner Valley and Hart Mountain in the distance

The soaking tub at Fisher Hot Springs

soaking tank needs to be filled with the 154-degree-F hot springs water before you can relax. Several hoses are embedded in the hot-water stream that tumbles down from the hot springs source about 50 yards to the west of the soaking tank. Take the end of one of the hoses out of the stream and prop it up at the east end of the tank to fill it. Before filling the tub, check for a wooden stopper in the drain at the bottom of the tank (you may need to devise your own plug if the stopper is missing). Fill the tub to a desired level (which can take up to a half hour). The water will probably be too hot to soak in immediately, so you may want to take a hike around the valley floor or explore the hot springs source while you wait for the tank to cool. When you're done soaking, drain the tank so it's ready for other soakers to fill. (Some soakers fill the tank halfway before they leave so that the next group of visitors have cooler water ready to mix with the fresh hot water from the hose.) In recent years some considerate soakers have brought in a large plastic tarp to line the steel tank—this reduces water leakage and provides a more comfortable sitting surface.

NORTHEAST OREGON

24. RITTER HOT SPRINGS

General description: A historic overnight stop on the old stagecoach road between Pendleton and John Day.

Location: Northeast Oregon, 90 miles south of Pendleton.

Development: The rustic resort has changed little from the turn of the century. It features an old hotel, a general store, and a warm-water swimming pool.

Best time to visit: Ritter Hot Springs is open Memorial Day to Labor Day; it is closed in the winter months. Resort hours are Sun through Wed from 8 a.m. to 10 p.m. and Thurs from 8 a.m. to 6 p.m. The resort is closed from sundown Fri to sundown Sat.

Restrictions: Swimsuits are required in the pool.

Access: Any vehicle can drive to the hot springs on the paved road that parallels the Middle Fork of the John Day River. (The resort is not open in winter, so there is little danger of snow-covered roads.)

Water temperature: The hot springs emerge from the ground at 106 degrees F, which is about the temperature of the water in the cinder-block soaking rooms. Hot water is also piped across the Middle Fork of the John Day River to the swimming pool, which averages 85 degrees F.

Nearby attractions: Anglers are often seen along the Middle Fork of the John Day River. Camping, hiking, and hunting are the main attractions in the Blue Mountains and the North Fork John Day Wilderness east of Tollbridge Campground.

Services: A few snacks and ice cream bars are available in the old Ritter Springs General Store. Pay for your snacks and your swimming pool fee under the honor system by leaving your cash in the box on a nearby table. The Ritter Springs Hotel features 8 rustic rooms with single and double beds. Two spacious cabins that sleep up to 7 guests are available across the creek from the Ritter Springs Hotel.

Camping: Camping and RV spots are available behind the Ritter Springs Hotel. Another overnight option is the Forest Service's Tollbridge Campground, 14 miles south of Ukiah on US 395, then 1 mile east on FR 10.

Maps: Oregon State Highway Map; *DeLorme: Oregon Atlas & Gazetteer,* page 77, A10.

GPS coordinates: N44.8916' / W119.1425'

Contact info: Ritter Hot Springs, Box 16, Ritter, OR 97872; (541) 421-3846; www.ritterhotsprings.com.

Finding the springs: From Ukiah head west on OR 244 for 1 mile to US 395. Drive south on US 395 for 39 miles to CR 15. (You'll come to the turnoff just before crossing a bridge over the Middle Fork of the John Day River.) Turn west onto CR 15. There may be a Ritter Hot Springs information sign near the turnoff, which will let you know if the resort is open. Follow CR 15 west parallel to the Middle Fork of the John Day River for 10 miles. Turn north onto a gravel road marked with a Ritter Hot Springs sign and drive about 1.5 miles. Park your car next to the Ritter Springs General Store. The swimming pool and hotel are both within 100 yards of the parking area.

THE HOT SPRINGS

Pulling up to the Ritter Springs General Store puts you in company with the thousands of cowboys and settlers who stopped here on the old stagecoach run to Pendleton. The Ritter Stage Road followed winding river valleys and steep ridges, culminating in a steep, 5-mile descent to the Ritter Springs stage stop. Frontier travelers surely welcomed the respite from the harrowing trip provided by the swimming pool and hotel at Ritter Springs.

Present-day visitors delight in the same mineral water and rustic lodging that stagecoach passengers enjoyed one hundred years ago. The 40-by-60-foot swimming pool is kept around 85 degrees F, a comfortable temperature for swimming during the May-to-October resort season. For a warmer soak take the footbridge across the

The warm-water swimming pool at Ritter Hot Springs

Middle Fork of the John Day River. A hundred yards up the hill from the riverbank are four private soaking rooms in a dilapidated cinder-block building (which is near the source of the hot springs). These soaking rooms can be filled with hot water to bring them up to a toasty 106 degrees F. Check with the staff (who usually are hanging around the general store) before using these soaking rooms to be sure they are available.

History

The hot springs were originally called McDuffie Springs, named for William Neal McDuffie, who discovered them in the early 1880s. Joseph Ritter, a Baptist minister who settled near the hot springs property, established the first post office in the region. The Ritter post office was later moved to the general store at the little resort, and the preacher's name became thereafter linked to the hot springs.

The Ritter Springs Hotel and General Store served as a stagecoach stop on the old Ritter Road between Pendleton and John Day. The general store was built in 1894. A faded sign above the entrance advertises what must have been considered essential supplies to travelers at the turn of the century:

The Ritter Springs Hotel, built in 1905

Simple rooms are available in the Ritter Springs Hotel, just steps from the warm-water swimming pool.

1894—Ritter Springs—1894

General Store & Stage Stop

Ranchers and Cowboys Supplies,

Bull Durham Chawin Tobacco,

45 Colt & Winchester Ammunition

First Aid Kits

The two-story Ritter Springs Hotel was built in 1905. The wooden clapboard building features eight guest rooms on the first floor. The second floor was used both as a meeting place and a dance hall.

Visitors from as far away as Portland made yearly pilgrimages to soak in the hot springs at Ritter. Local historian Jo Southworth recalled the resort's popularity in a 1972 article in the *Blue Mountain Eagle:*

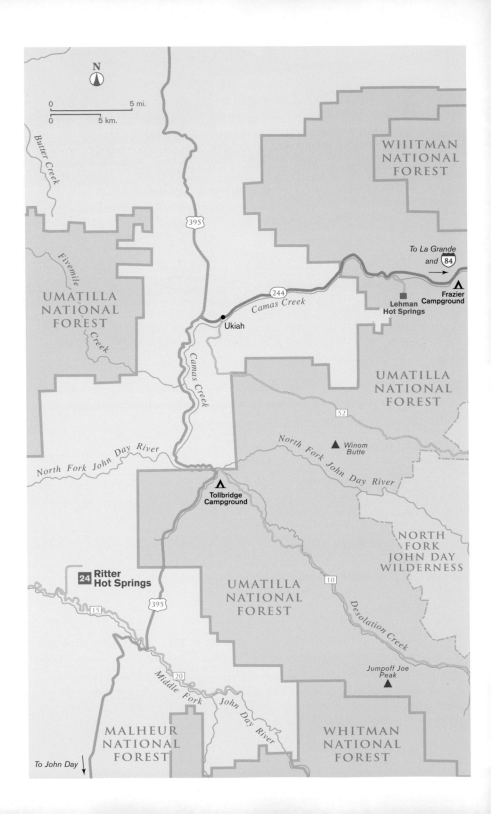

N

0 ____ 5 mi.
0 ____ 5 km.

Butter Creek

WHITMAN
NATIONAL
FOREST

To La Grande
and 84

Fivemile

[395]

UMATILLA
NATIONAL
FOREST

Creek

[244]

Camas Creek

Ukiah

■ Lehman
Hot Springs

⚑ Frazier
Campground

UMATILLA
NATIONAL
FOREST

[52]

Camas Creek

▲ Winom
Butte

North Fork John Day River

North Fork John Day River

⚑ Tollbridge
Campground

NORTH
FORK
JOHN DAY
WILDERNESS

24 Ritter
Hot Springs

[15]

UMATILLA
NATIONAL
FOREST

[10]

Desolation Creek

[395]

[20]

Middle Fork

John Day River

Jumpoff Joe
Peak

▲

MALHEUR
NATIONAL
FOREST

WHITMAN
NATIONAL
FOREST

To John Day ↓

The Ritter Hot Springs General Store and Stage Stop, built in 1894

In spite of the challenge of reaching Ritter, people went for a day or for weeks. They stayed in the hotel or the cabins, or camped in their own tents. Victims of all types of rheumatism, skin disease, stomach trouble and other ailments sought relief from their afflictions. Some of the inflammatory rheumatism cases came in on stretchers. The grayish green moss that grew in the mineral water helped skin infections when bandaged to the sores. People dried the moss and took it home for further treatment. Stomach patients drank the water. One of them was so miserable he could hardly eat. He was restored to such good health by Ritter Hot Springs treatment that he even enjoyed eating spare ribs with the other diners by the end of the day.

25. HOT LAKE SPRINGS

General description: A beautifully restored hotel and resort on the shores of a warm-water lake.

Location: Northeast Oregon, 8 miles southeast of La Grande in the Grande Ronde Valley.

Development: Hot Lake Springs has been commercially developed since the 1860s. The massive brick hotel lay in ruins for decades, until David Manuel purchased the hotel in 2003 and spent nearly a decade bringing it back to its historical elegance. In 2020 new owners Mike and Tamarah Rysavy began a major upgrade to the hotel, which will include several new soaking venues, intimate pubs scattered throughout the building, and a small movie theater, among other improvements.

Best time to visit: Hot Lake Springs is open year-round. Phone ahead for reservations during popular holiday weekends.

Restrictions: Swimsuits are required in public bathing areas.

Access: Any vehicle can make the trip on the paved highway to the hotel.

Water temperature: The hot springs are located in a small building between the lake and the hotel. The artesian hot springs produce 2.5 million gallons per day of 198-degree-F water.

Nearby attractions: Ladd Marsh, a 3,000-acre wildlife sanctuary, is adjacent to Hot Lake. It's a great place to spend a few hours observing dozens of waterfowl species. There's a nature trail through the sanctuary. Hot Lake Springs is on the 95-mile-long, figure-eight Grande Tour from La Grande past Hot Lake Springs and on to Union and Medical Springs. From Medical Springs the driving tour then heads back to Union and north to Cove before returning to La Grande. This is a nice daylong tour if you're staying in La Grande or at Hot Lake Springs. Pick up a tour map and area highlight descriptions at the La Grande/Union County Visitors and Convention Bureau.

Services: The refurbished resort features 14 guest rooms with period furniture, private soaking tubs for overnight guests, a restaurant, a museum with a large collection of Native American artifacts, a bronze foundry, public outdoor soaking tubs, and a spa. Room reservations are taken by phone only. David Manuel, an internationally known bronze sculptor, creates and displays his works at the hotel, and art collectors come from around the world to buy his latest bronze creations.

Camping: Grande Hot Springs RV Resort, about a half mile northwest of Hot Lake Springs, has more than 100 spaces for motor homes and a large camping area for tents.

Maps: Oregon State Highway Map; *DeLorme: Oregon Atlas & Gazetteer,* page 75, E6.

GPS coordinates: N45.2459' / W117.9582'

Contact info: Hot Lake Springs, 66172 Hwy. 203, La Grande, OR 97850; (541) 963-4685; https://www.hotlakelodge .com; info@hotlakelodge.com

Finding the springs: From La Grande drive south on I-84 for 2.5 miles to exit 265. Take the exit and drive east for 5 miles on OR 203. On the right-hand side of the highway, you'll see the warm-water lake and elegant brick hotel. Drive through the entrance gates and up to the front of the hotel.

The historic Hot Lake Springs Hotel

History

Hot Lake was likely the first hot springs in the Pacific Northwest to have been seen by Europeans. Robert Stuart, a member of the Astor Expedition, was returning from the Oregon Coast to St. Louis, Missouri, in the summer of 1812. Stuart's observations were later recorded by writer Washington Irving:

Emerging from the chain of Blue Mountains, they descended upon a vast plain, almost a dead level, sixty miles in circumference, of excellent soil, with fine streams meandering through it in every direction, their courses marked out in the wide landscape by serpentine lines of cotton-wood trees, and willows, which fringed their banks, and afforded sustenance to great numbers of beavers and otters.

In traversing this plain, they passed, close to the skirts of the hills, a great pool of water, three hundred yards in circumference, fed by a sulphur spring, about ten feet in diameter, boiling up in one corner. The vapor from this pool was extremely noisome, and tainted the air for a considerable distance. The place was much frequented by elk, which were found in considerable numbers in the adjacent mountains, and their horns, shed in the spring-time, were strewed in every direction around the pond.

—from Astoria; or, Anecdotes of an Enterprise
Beyond the Rocky Mountains (1836)

A few decades after Stuart's visit, pioneers trekked across the Oregon Trail on their way west. The wagon trail passed within sight of Hot Lake, and the pioneers often stopped to rest, wash their clothes, and bathe in the hot water.

Eventually the hot springs and adjacent warm-water lake were incorporated into a surrounding cattle ranch. In 1914 the *Oregonian* recounted a story of one of the early ranch hands who was enamored with the healing properties of Hot Lake:

> [At one time] Hot Lake was merely part of a ranch occupied by a former sailor named Tommy Atkins. He was most loyal to the curative power of healing Hot Lake, whose boiling, steaming waters had for centuries been the faithful remedy for hosts of Indians from near and far. Tommy swore loudly that the hot water of this odd lake would absolutely cure any human ailment.
>
> Tommy was the proud owner of 30 horses, among which was a young mare, beautiful in form and color, but wild, fiery and unbroken. Tommy swore that he would tame and ride this animal. The result of the first lesson was that the indignant equine pupil hoisted aloft the sailor, then stamped, kicked and bit him. The maimed Tommy was picked up for dead and hauled to his beloved Hot Lake.
>
> He balked at any doctoring, swearing fluently at the medical men's verdict that he had three cracked ribs, a broken arm, fractured leg and sundry serious internal injuries. After nightfall, when the doctors had given up hope and departed, the apparently doomed sailor quietly crawled out to the shore of the lake. With a volley of muttered oaths, he flopped into the steaming water and floundered around like a fish or lobster.
>
> By chance he drifted too near to where the boiling, fiery water laps up from the earth and is 15 feet deep. Too weak to resist, Tommy was drawn into this hissing whirlpool of Hades. His howls, yells, yowls, whoops and gurgling curses brought a neighboring rancher out on the run. The rancher heaved a rope to the flopping, screeching victim and towed the partly cooked, rosy red sailor to the harbor of solid land. Some reports assert that this Tommy Atkins was cured of every ailment, including his swearing habit.

In 1864 a small hotel was built on the property. The hotel was heated with water from the hot springs, one of the first commercial uses of geothermal energy in the United States. Dr. H. J. Minthorn leased the property in 1889 and added two wings to the little hotel. Other new features included a new bathhouse and an octagonal springhouse that enclosed the bubbling hot water.

The resort was expanded again in 1908 with the addition of a 200-foot-long brick hospital with 105 rooms. The new hospital could provide therapeutic mineral-water soaks to more than 200 patients a day in large tiled rooms containing rows of sunken bathtubs. There were also a surgical area (complete with elevated observation area) and a dining hall that could seat 1,500. At its height the hospital was staffed with four physicians, fifteen nurses, an X-ray technician, and a bacteriologist. Other additions soon followed, including a barbershop, a ballroom, a poolroom, bathhouses, a

The original bathhouse on the shores of Hot Lake at Hot Lake Springs

drugstore, a cafeteria, and a hospital. The sheer size and variety of facilities located in the building earned the resort the label "The Town Under One Roof."

The Hot Lake Hospital and Sanitarium was also known as "The Mayo Clinic of the West." A promotional flyer from the regional railroad company extolled the curative properties of the natural hot springs:

The largest, hottest, and most curative springs known; best bathing facilities, most courteous attendants; first-class medical and surgical conveniences; finest operating room in the west; steam heat, electric lights; hot and cold water throughout the building.

Another advertisement touted the benefits that guests received from bathing in and drinking the hot mineral springs:

The water, pleasing to the taste, has cured and restored to health innumerable invalids, who had tried in vain much-advertised and noted resorts. The treatments consists of copious drinking of the water, hot-water baths, hot-vapor baths and hot-mud baths—the heat in all cases being from the water as it comes from the ground. A poultice from the sediment at the bottom of the lake relieves the most agonizing form of rheumatism, and reduces the

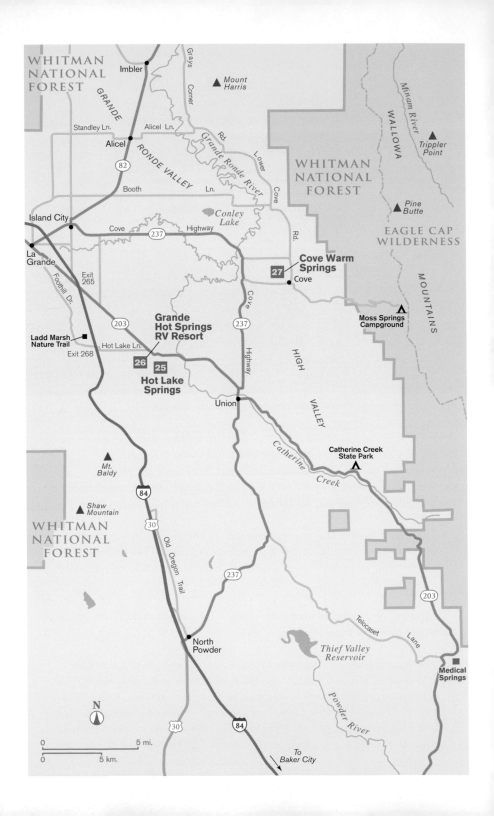

WHITMAN
NATIONAL
FOREST

Imbler

Grays
Corner

▲ Mount
Harris

GRANDE

Standley Ln.   Alicel Ln.

Alicel

Grande Ronde River

Rd.

Lower

Minam River

WALLOWA

▲ Trippler
Point

82

RONDE VALLEY

WHITMAN
NATIONAL
FOREST

Booth   Ln.

Cove

▲ Pine
Butte

Island City

Cove

Conley
Lake

Highway

237

Rd.

EAGLE CAP
WILDERNESS

La
Grande

Exit
265

Cove Warm
Springs

27

Cove

MOUNTAINS

Foothill Dr.

203

Grande
Hot Springs
RV Resort

237

Cove

Moss Springs
Campground

Ladd Marsh
Nature Trail

Hot Lake Ln.

26 25

Highway

HIGH

Exit 268

Hot Lake
Springs

Union

VALLEY

▲ Mt.
Baldy

Catherine

Catherine Creek
State Park

Creek

84

30

▲ Shaw
Mountain

Old Oregon Trail

WHITMAN
NATIONAL
FOREST

237

203

Telocaset

Lane

North
Powder

Thief Valley
Reservoir

Medical
Springs

N

Powder River

0   5 mi.

0   5 km.

30

84

To
Baker City

The entrance to the grounds of the Hot Lake Springs Hotel

swelling. Long-standing cases of dyspepsia have been cured by a few weeks' use of the water.

On May 7, 1934, a fire destroyed all the wooden buildings on the property, leaving only the brick hotel. The fireproof building was reopened after the fire, but the resort's popularity never regained the levels seen before the fire. The property changed hands several times over the next fifty years, finally serving as a retirement home in the 1970s. A string of owners in the 1980s and 1990s proposed grand ideas to revitalize the resort to its former glory, but none of these plans were realized, and the old hotel became a vandalized eyesore.

The darkened hotel also became the source of several ghost stories. A caretaker in the 1990s reported hearing a piano playing on the third floor, even though the building was empty. He also heard ghostly screams coming from the old surgery room and observed that three rocking chairs sitting in a row on the third floor "never seem to get dusty," as if somebody or something was sitting in them.

In 2003 the resort's luck changed when David and Lee Manuel purchased the hotel and surrounding grounds. Over the next decade they refurbished the old resort, which reopened to the public in 2010. Hot Lake Hotel was purchased by Mike and Tamarah Rysavy in the spring of 2020. The Rysavys also own the adjacent Grande Hot Springs RV Resort. The Rysavys intend to rename the resort to The Lodge at Hot Lake Springs.

26. GRANDE HOT SPRINGS RV RESORT

See map on page 124.

General description: A pleasant overnight spot for the RV or camping crowd, with a quiet tree-lined stream, picnic spot, and outdoor soaking pools.

Location: Northeast Oregon, 8 miles southeast of La Grande in the Grande Ronde Valley.

Development: The RV park and pools have been developed for decades. Extensive upgrades, including private soaking tubs and overnight yurt accommodations, have been added by the owners.

Best time to visit: The RV park is open year-round. Spring, summer, and fall are the busiest times of year.

Restrictions: The mineral-water soaking pools are open from 9 a.m. to 8 p.m. The pools are available to registered resort guests only. Currently the pools are seasonal (closed in the winter). Call ahead in early spring or late fall to see if the pools are open.

Access: Any vehicle can make the trip.

Water temperature: Grande Hot Springs RV Resort features 2 heated soaking pools behind the main lodge. Both pools are supplied with hot water heated via a heat exchanger connected to a 170-degree-F mineral-water well. Both pools are maintained at around 100 degrees F.

Nearby attractions: Ladd Marsh, a 3,000-acre wildlife sanctuary adjacent to the RV resort, is a great place to spend a few hours observing dozens of waterfowl species. There's a mile-long nature trail through the sanctuary.

Grande Hot Springs RV Resort is on the 95-mile-long, figure-eight Grande Tour from La Grande past Hot Lake Springs and on to Union and Medical Springs. From Medical Springs the driving tour then heads back to Union and north to Cove before returning to La Grande. This is a nice daylong tour if you're staying anywhere in the area. Pick up a tour map and area highlight descriptions at the La Grande/Union County Visitors and Convention Bureau.

Also consider taking a tour of the neighboring Hot Lake Springs hotel, including the bronze foundry and gallery of artist David Manuel.

Services: The RV resort features a small restaurant/snack bar, a limited selection of groceries, a 24-hour laundry room, a picnic area, and a shower building. The owners of the RV park, Mike and Tamarah Rysavy, have added several yurts, cabins, and private outdoor mineral-water soaking tubs since purchasing the property in 2013.

Camping: More than 100 full-service RV pull-thrus are available, as well as a separate creek-side tent camping area shaded by willow and cottonwood trees. More upscale accommodations include a 27-foot Airstream Land Yacht (built in 1970). The Pendleton-themed Airstream trailer sleeps up to 3 guests. Also available are 2 yurts with queen beds, private soaking tubs, and private picnic tables and fire pits.

Maps: Oregon State Highway Map; *DeLorme: Oregon Atlas & Gazetteer,* page 75, E6.

GPS coordinates: N45.2456' / W117.9692'

Contact info: Grande Hot Springs RV Resort, 65182 Hot Lake Ln., La Grande, OR 97850; (541) 963-5253; www.grandehotsprings.com; info@grandehotsprings.com.

Finding the springs: From La Grande drive south on I-84 for 2.5 miles to exit 265. Take the exit and drive east for 5 miles on OR 203. Turn right (east) onto Hot Lake Lane (a few yards before the Hot Lake Springs hotel entrance). Drive 0.5 mile on this gravel road to the Eagles Hot Lake RV Park—you'll see a sign at the entrance on the left-hand side of the road.

THE HOT SPRINGS

Located less than half a mile from the refurbished Hot Springs Lake Hotel, the 30-acre Grande Hot Springs RV Resort reopened under new management in 2013. The resort has been a well-known stop for motor-home tourists passing through the Grande Ronde Valley for decades.

The main lodge contains a small grocery store, a laundry area, and showers. The walls of the lodge are lined with large black-and-white photos of the nearby Hot Lake Springs hotel during its heyday in the 1920s.

The public hot-water soaking pool at Grande Hot Springs RV Resort MICHAEL RYSAVY

The welcome sign at Grande Hot Springs RV Resort MICHAEL RYSAVY

A private soaking tub adjoins a yurt at Grande Hot Springs RV Resort. MICHAEL RYSAVY

Behind the lodge are two hot-water soaking pools. Water from a 170-degree-F well flowing at 40 gallons per minute is run through a heat exchanger to raise the temperature of the pools. The largest soaking pool is 30 feet in diameter and 3 feet deep. The second pool measures 6 feet by 12 feet and is 2 to 3 feet deep. The hot soaking pool is kept between 102 and 104 degrees F, while the cooler soaking pool is kept at 92 to 96 degrees F.

The hot water also provides space heat for the lodge, office, pavilion, and an apartment.

Additional improvements include a private outdoor soaking area with clawfoot tubs, as well as a communal soaking area with a large soaking tub. Ten yurts are also on the drawing board, each with a private soaking area planned near the tree-lined creek.

27. COVE WARM SPRINGS

See map on page 124.

General description: A natural warm-water swimming pool and picnic area on the edge of a small town in Oregon's Grande Ronde Valley.

Location: Northeast Oregon, 17 miles east of La Grande in the community of Cove.

Development: A concrete swimming pool encloses the natural warm springs. A small shaded picnic area is adjacent to the pool.

Best time to visit: Late May to Labor Day. The swimming pool is closed other times of the year.

Restrictions: Swimsuits are required in the pool, and a small fee is charged to swim.

Access: Any vehicle can make the trip on the paved roads to Cove. The pool is open 7 days a week from 11 a.m. to 8 p.m. Memorial Day to Labor Day.

Water temperature: The warm springs and the swimming pool are both 86 degrees F.

Nearby attractions: The charming Victorian-era town of Union is located about 7 miles south of Cove on OR 237. The Union County Museum features excellent cowboy and geology exhibits, as well as period rooms. The Eagle Cap Wilderness in the Wallowa Mountains east of Cove offers hiking and fishing.

Services: There are a few snacks for sale at the pool. Gas, lodging, and more substantial meals are available in the nearby towns of La Grande and Union. A cabin is available for overnight rental on the Cove Warm Springs property. A more upscale lodging option in the nearby town of Union is the Union Hotel, built in 1921 in the American Renaissance style (call 541-562-6135 or visit www.theunionhotel.com).

Camping: Tent camping and RV parking are available on the 25-acre property surrounding the warm springs. Another option is the Moss Springs Campground in the Eagle Cap Wilderness east of Cove. (Take French Street in Cove east until it bends into Mill Creek Lane, which soon becomes FR 6220. This road climbs for 8 miles to the Moss Springs Campground.) At a lower elevation is Catherine Creek State Park, 8 miles southeast of Union on OR 203.

Maps: Oregon State Highway Map; *DeLorme: Oregon Atlas & Gazetteer,* page 75, E7.

GPS coordinates: N45.2935' / W117.8071'

Contact info: Cove Warm Springs, 907 Water St., Cove, OR 97824; (541) 568-4890; http://coveoregon.org/cove-warm-spring-pool/. Check their Facebook page for the latest information, www.facebook.com/CoveSwimmingPool.

Finding the springs: From La Grande drive 17 miles east on OR 237 to the town of Cove. Turn left off Main Street onto French Street and follow the Cove Swimming Pool signs to the pool parking area.

THE HOT SPRINGS

Although Cove Warm Springs isn't a steaming wilderness soak, it is a nice place for a family swim and picnic on a summer afternoon. Unlike many pools where mineral water is piped in, the Cove swimming pool is built directly on top of the warm springs. Most of the pool bottom is concrete, but a 15-by-15-foot opening in the center of the pool exposes bare rock fissures that gush warm water into the pool at more than 300 gallons per minute. The swimming pool isn't chlorinated, since the large flow rate of the warm springs ensures that the water in the pool turns over several times a day, which keeps state water-quality regulators at bay. A picnic area with big cottonwood trees next to the pool provides a relaxing location for a late lunch after an hour or two of swimming.

History

A nearby church camp often sends young people to swim in the pool on hot summer days. According to the pool owner, South Africa's Archbishop Desmond Tutu visited the church camp several years ago and spent an afternoon swimming with the kids in Cove Warm Springs. The swimming pool has had the same owner since the 1980s. The owner lives in an adjacent house but closes the pool in fall.

The swimming pool at Cove Warm Springs

SOUTHEAST OREGON

28. CRANE HOT SPRINGS

General description: A small rustic resort with an open-air hot pool and enclosed soaking tubs, situated on windswept sagebrush flats in eastern Oregon.

Location: Southeast Oregon, about 25 miles southeast of Burns.

Development: This private resort has had a series of uses over the past 100 years.

Best time to visit: Autumn, winter, and spring are best for soaking in the indoor soaking tubs. The cooler outdoor pool is a favorite in the summer months. According to the present owners, the resort keeps busy year-round, but they rarely have to turn people away. Hunters, schoolchildren from the nearby town of Crane, and tourists heading toward Steens Mountain and the Alvord Desert are the resort's main customers.

Restrictions: The outdoor soaking pool and the indoor soaking tubs are open from 9 a.m. to 9 p.m. every day. Bathing suits must be worn in the open-air hot springs pond. What you wear in the private enclosed hot tubs is up to you. A fee is charged for soaking and overnight accommodations.

Access: Any vehicle can make the trip along OR 78 to the hot springs.

Water temperature: Hot water is pumped from 2 wells on the property, one around 180 degrees F and a second cooler well at 120 degrees F. Water from a cold-water well is mixed with the hot well water in the soaking pond and hot tubs. The outdoor hot springs pond varies seasonally between 90 and 102 degrees F. The private soaking tubs can be adjusted to any desired soaking temperature.

Nearby attractions: The 183,000-acre Malheur National Wildlife Refuge, situated about 25 miles from Crane Hot Springs (south of Burns on OR 205), is a major resting area for migrating birds on the Pacific Flyway. More than 250 species of birds have been identified on the refuge. Malheur Cave is located 17 miles east of the hot springs. Stretching more than 3,000 feet in length, the cave is really a lava tube, formed thousands of years ago when volcanic lava solidified around a fast-moving river of molten rock. When the interior lava spilled out of the tube, the solid rock walls of the tube remained, forming the present cave. Another geologist's delight is Diamond Craters, located 25 miles south of the hot springs. Diamond Craters contain an outstanding selection of volcanic formations, including lava tubes and cinder cones.

Services: Few supplies are available at the hot springs—stop at Crane Supply, about 3 miles southeast of the hot springs in the little town of Crane, and stock up (either there or in Burns) on all the food, drinking water, and other supplies you'll need if you're staying a few days at the hot springs. Crane has a wide variety of lodging options, including 9 cabins, 9 RV spots, a 3-bedroom ranch house, a 3-bedroom inn, and a seasonal tepee that contains a private mineral-water hot tub. The Crane Creek Suites, completed in 2017, includes 5 brand-new guest rooms. Each room includes an en-suite bathroom and private patio with a heated floor and soaking tub. Along with your stay, enjoy unlimited use of the hot springs pond and all facilities.

Camping: The resort offers 15 tent spots for overnight guests. The resort also offers upscale "glamping," where guests stay in tepees complete with queen beds, fire pit, and individual soaking tubs. A genuine sheepherder's wagon has also been converted into a unique overnight accommodation for those looking for a quirky place to lay their heads.

Map: *DeLorme: Oregon Atlas & Gazetteer,* page 82, D2.

GPS coordinates: N43.4404' / W118.6289'

Contact info: Crane Hot Springs, 59315 Hwy. 78, Burns, OR 97720; (541) 493-2312; www.cranehotsprings .com.

Finding the springs: From Burns drive 25.5 miles southeast on OR 78. Look for the Crane Hot Springs sign on the north side of the highway (between mileposts 25 and 26). Turn left and drive about 0.1 mile to the hot springs parking lot. The small town of Crane is located about 3.5 miles southeast of the hot springs.

Galvanized stock tanks make appropriate soaking tubs at Crane Hot Springs. DEBRA KRYGER

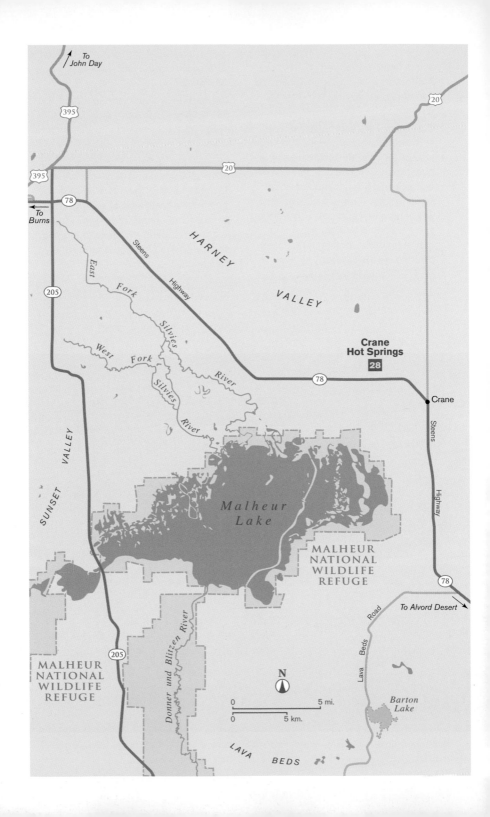

To
John Day

395

20

395

78

To
Burns

205

H A R N E Y

East Fork

Steens

Highway

V A L L E Y

West Fork

Silvies

Silvies River

River

**Crane
Hot Springs**
28

78

Crane

Steens

SUNSET VALLEY

*Malheur
Lake*

Highway

MALHEUR
NATIONAL
WILDLIFE
REFUGE

78

To Alvord Desert

Lava Beds Road

MALHEUR
NATIONAL
WILDLIFE
REFUGE

205

Donner und Blitzen River

N

0   5 mi.

0   5 km.

*Barton
Lake*

L A V A   B E D S

THE HOT SPRINGS

The large open-air soaking pond is about 80 feet in diameter, gently graduating in depth from its grassy edges to more than 12 feet in its center. A wooden platform extends a few feet in from the edge of the pool. Picnic tables are available near the pond (bring your own lunch).

The bathhouse adjacent to the soaking pond contains six individual soaking tubs. The tubs are made from galvanized livestock watering tanks, which provide a truly unique soaking experience.

The private, wood-paneled tub rooms are rented by the hour, and water temperature can be adjusted to as hot as you can stand it. Four of the soaking tubs are oblong (about 3 feet by 10 feet) and can easily hold two to four people. Two larger circular tubs are available for bigger groups or families. All soaking tubs are about 3 feet deep.

History

Three natural hot springs flowed onto the sagebrush prairie for thousands of years before European settlers moved into the surrounding Harney Valley. In the early 1900s women from the nearby community of Crane sometimes did their laundry in

A variety of lodging options are available at Crane Hot Springs. DEBRA KRYGER

An indoor soaking tub at Crane Hot Springs

The outdoor soaking pool at Crane Hot Springs

the hot springs. They strung their clotheslines near the springs so that the clean, wet clothes could dry in the high desert breezes. One innovative housewife at the turn of the century transported a hand-operated washing machine to the hot springs to speed up the laundry day chores. Local ranchers also took advantage of the commercial benefits of the hot water. The ranchers would drive truckloads of butchered hogs to the property, where they dipped the carcasses into the scalding water to loosen the animal hides.

In the 1920s the hot springs acquired a more sophisticated use. Minne Iland, a local physician, formed a partnership with entrepreneur Ralph Catterson and built a 30-by-60-foot swimming pool near the springs, piping the 180-degree-F artesian hot water into the pool. No cold water was yet available at the little resort, so bathers had to wait a day or more for the water to cool enough to their liking. Iland and Catterson also built a dance hall and a restaurant near the swimming pool. Taking the last initials of their names, Iland and Catterson christened their little resort "I and C Hot Springs." The wooden dance hall and restaurant were destroyed in a fire in the 1930s and never rebuilt. The resort name was later changed to Crystal Crane Hot Springs, perhaps after nearby Crane, Oregon (which itself may have been named for the sandhill cranes that were once abundant in the area). In 2019 the name was shortened to simply Crane Hot Springs.

From the 1930s until the 1990s, a series of entrepreneurs managed the hot springs. One local named Shorty Lasater opened a gas station and grocery store near the springs in the 1930s, but the business survived only a few years. Now the only remnant of these many commercial ventures is a decaying concrete swimming pool located behind the current lodge.

Tepee lodging at Crane Hot Springs DEBRA KRYGER

The latest round of renovations to Crane Hot Springs is due to the hard work of Dan and Denise Kryger, who purchased the property in 1997. The Krygers are responsible for many improvements at the resort, including the hot tub rooms, tepees, cabins, and upgraded RV facilities.

29. JUNTURA HOT SPRINGS

General description: A secluded hot springs pool located in the center of an island in the Malheur River, often too hot for comfortable soaking.

Location: Southeast Oregon, 60 miles east of Burns near the town of Juntura.

Development: None.

Best time to visit: The winter months may be the best time to visit Juntura Hot Springs. The summertime temperature in the soaking pools can easily exceed 110 degrees F, but on a cold winter day, the hot-water soak is more tolerable. It's also essential to visit the hot springs when the water level in the Malheur River is low. The crossing between the riverbank and the island can be treacherous when the river is high from spring runoff. The autumn months feature beautiful fall colors along the river, but the hot springs are often overrun with hunters who use the hot springs and surrounding area as their camping base.

Restrictions: No restrictions; nudity is common.

Access: Any vehicle can make the trip on the paved highway from the town of Juntura to the old highway bridge. You'll probably need to hike the last 0.5 mile to the hot springs and then face what can be a difficult crossing across the Malheur River to reach the island that's home to the hot springs pools.

Water temperature: The main soaking pool containing the hot springs can reach a toasty 115 degrees F (although this can cool off a few degrees in winter). The shallower soaking pools that take the runoff water from the main pool range from 90 to 105 degrees F, depending on the amount of river water that mixes in.

Nearby attractions: The 183,000-acre Malheur National Wildlife Refuge, situated about 70 miles southwest of Juntura, is a major resting area for migrating birds on the Pacific Flyway. More than 250 species of birds have been identified on the refuge. Malheur Cave is located about 80 miles south of Juntura. Stretching more than 3,000 feet in length, the cave is really a lava tube, formed thousands of years ago when volcanic lava solidified around a fast-moving river of molten rock. When the interior lava spilled out of the tube, the solid rock walls of the tube remained, forming the present cave. Another geologist's delight is Diamond Craters, which contains an outstanding selection of volcanic formations, including lava tubes and cinder cones.

Services: None available at the springs. The Oasis Café, located in Juntura 3 miles west of Juntura Hot Springs on US 20, is well known for its hamburgers and milkshakes. It's also a good place to stock up on water before visiting the hot springs. A small motel is connected to the cafe if you need overnight accommodations. An RV park next to the Oasis Café fills with hunters in the fall.

Camping: You can camp anywhere on BLM land near the hot springs or on the surrounding riverbanks. A more developed campground is available at Warm Springs Reservoir, located 9 miles southeast of the town of Juntura.

Maps: Oregon State Highway Map; *DeLorme: Oregon Atlas & Gazetteer,* page 82, B5.

GPS coordinates: N43.7759' / W118.0478'

Contact info: Bureau of Land Management, Vale District Office, 100 Oregon St., Vale, OR 97918; (541) 473-3144; www.blm.gov/office/vale-district-office.

Finding the springs: From the town of Juntura, head east on US 20 for 2.4 miles. Make a sharp left off US 20 just before crossing a highway bridge (you'll be turning onto a section of old highway). Proceed about 0.1 mile to the original old bridge that crossed the highway. Park your car before crossing this old bridge and proceed on foot. Cross the old bridge and immediately turn left, then follow the dirt road that hugs the bank of the river for 0.5 mile. As you come around a sharp bend in the river, you'll pass some bare campsites and a fire pit or two. Immediately after these campsites, the road dips into the river and

emerges about 10 feet away on the island containing Juntura Hot Springs. If it's a very dry time of year, you may be able to easily wade the river to the island, but the water level can be dangerously high in the springtime, and the crossing may not be possible. Once you cross onto the small island, you'll see the main hot springs on the island's highest point. Cooler soaking pools are located on the river bank about 10 feet from the main pool.

(The old bridge crossing the highway may be barricaded with cement barriers, and vehicles won't be able to proceed past this point. If the barricades are down, it may be possible to drive on the dirt road around the bend, and even to cross the ford and drive right up to the hot springs on the island. But it may be best to leave your vehicle at the old bridge, even if you could drive on the road, and enjoy the walk.)

The main soaking pool at Juntura, with concrete slab on the edge

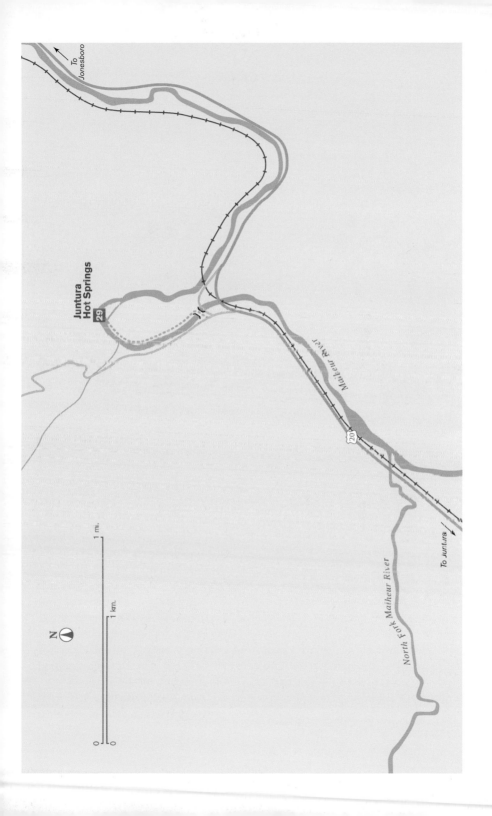

The runoff pools at the edge of the island at Juntura Hot Springs

THE HOT SPRINGS

Juntura Hot Springs is also called "Horseshoe Bend Hot Springs," named after the sharp bend in the river just before you reach the island where the hot springs emerge. The main pool is in a beautiful location in the center of the island on the Malheur River, but the 108-degree-F temperature (or higher) means that most soakers only stay in for a half hour or less. Visitors from November to early March probably have the best soaks, when the pool water temperature drops a few degrees. Try soaking in the riverside pools that collect runoff from the main island pool—they are often more comfortable.

30. MICKEY HOT SPRINGS

General description: A slumbering series of thermal features that during wet years may transform into a display of mud pots, hissing steam vents, and Oregon's only natural geyser.

Location: Southeast Oregon, 103 miles southeast of Burns in the Alvord Desert.

Development: The hot springs and adjacent dry desert lake beds are undeveloped, except for the Bureau of Land Management (BLM) fence that encloses the 20 acres surrounding the springs.

Best time to visit: You'll have the best chance of seeing gurgling mud pots during spring in years when there has been lots of rainfall (which doesn't happen all that often). The thermal activity is much quieter during normal years, when little rain falls. (After all, that's why this is a desert!) Try to miss the scorching temperatures of July and Aug. Roads in the area can turn into impassable mud during rainstorms, so keep an eye on the weather.

Restrictions: The BLM doesn't encourage soaking at Mickey Hot Springs due to the high temperatures and fragile nature of some of the geothermal formations. Signs are posted warning of the scalding temperatures in the thermal pools.

Access: The paved and graveled Fields-Denio Road is accessible by most vehicles. The final 6 miles to the hot springs is a heavily washboarded dirt road. During the rare rainstorms in the Alvord Desert, this road may be too muddy to drive.

Water temperature: BLM researchers have recorded temperatures as high as 206 degrees F in the thermal pools at Mickey Hot Springs. Temperatures vary from year to year and season to season. The temperature in the large main pool averages 130 degrees F—definitely too hot for soaking.

Nearby attractions: Few visitors make Mickey Hot Springs their sole destination on a trip to the Alvord Desert. Combine your search for Oregon's elusive natural geyser with a trip to Borax Lake and a soak in Alvord Hot Springs, and then head to the cooler high country of Steens Mountain to the west.

Services: No services are available. Be sure you have plenty of water and a full tank of gas before heading for the hot springs—it's a good hour's drive south to the nearest store and gas station in Fields. There's also a small motel and restaurant (with great milkshakes) in Fields. Call ahead to see if they have rooms available (541-495-2275).

Camping: Camping isn't allowed within the fenced area enclosing the hot springs, but you can camp in the parking area or on most of the thousands of surrounding acres of BLM land. Camping permits are not required. For more picturesque mountain camping, take one of the side roads that wander into the foothills of Steens Mountain west of Fields-Denio Road and camp anywhere on BLM land. (Check the BLM map to ensure you are on public property.) Camping is also available at the BLM's Mann Lake Recreation Site, 7 miles north on Fields-Denio Road from the turnoff to Mickey Hot Springs. No drinking water is available at Mann Lake, so be sure you have plenty with you. The 276-acre lake is popular with anglers in search of high-desert cutthroat trout.

Maps: BLM Steens High Desert Country map; *DeLorme: Oregon Atlas & Gazetteer,* page 86, B4.

GPS coordinates: N42.6785' / W118.3482'

Contact info: Bureau of Land Management, Burns District Office, 28910 Hwy. 20 W., Hines, OR 97738; (541) 573-4400; www.blm.gov/visit/mickey-hot-springs.

Finding the springs: From Burns drive 65 miles southeast on OR 78. Turn south onto the gravel Fields-Denio Road (also called Fields-Follyfarm Road) and drive 31.3 miles toward Fields. Turn east onto a dirt road just north of a cattle guard. Drive 6.4 miles east, skirting the southern tip of Mickey Butte, until you arrive at the small parking area near a BLM sign that warns of the boiling water in Mickey Hot Springs. Park your car near the sign and walk through the narrow gate. Follow a well-marked path about 200 feet east to the main hot pool at Mickey Hot Springs. Take along a copy of the BLM's Steens High Desert Country map of the area—the side roads can be confusing.

The large, hot pool at Mickey Hot Springs

One of the hot pools at Mickey Hot Springs

THE HOT SPRINGS

Mickey Hot Springs isn't just one hot spring—it's a group of close to two dozen steam vents, hot pools, mud pots, and occasionally a natural geyser. What thermal features you see depends on the time of year you visit and the amount of precipitation that has fallen recently. During extremely wet years the Mickey Hot Springs area is alive with hissing steam vents and gurgling thermal pools. Some of the steaming vents turn into soupy mud pots, bubbling with opaque minerals. The area is the only known location in Oregon that has a natural geyser, although it is rarely seen. In the late 1980s and early 1990s, the geyser was a churning, bubbling pool of 206-degree-F water and steam that sporadically erupted about a foot into the air. The geyser was even more spectacular during the wet spring of 1992, erupting 6 to 8 feet in the air every 1 or 2 minutes.

Unfortunately the geysers and mud pots are rarely seen. Most of the time the geothermal area is fairly quiet, with only a few bubbles in the main pool and some wisps of steam hissing in rock fissures to indicate thermal activity. The largest pool (called the "Morning Glory" by BLM personnel, after the famous thermal pool in Yellowstone National Park) is 20 feet in diameter and more than 10 feet deep. About 10 yards north of the main pool is what appears to be an extinct geyser cone, which has built up over thousands of years to rise more than 8 feet above the surrounding desert. Overflow from the Morning Glory pool forms a small stream that flows through a rock channel to a smaller pool about 10 yards to the south. This second pool is about 4 feet wide, 10 feet long, and 3 feet deep. The pool temperature in the overflow pool is around 120 degrees F. The water from this pool flows through another channel toward the flat desert playas, where it pools in cooler and shallower basins. Some visitors attempt to soak in the hot pools at the southern end of the channels, but these pools can still be too hot for comfort. Use extreme caution if you decide to try soaking at Mickey Hot Springs, and keep in mind that the BLM strongly discourages bathing here due to the fragile nature of the area and the wide temperature fluctuations that occur in the thermal pools.

History

For decades the Alvord Ranch held a BLM lease to graze cattle on rangeland surrounding Mickey Hot Springs. Every year a few diehard hot springs enthusiasts would visit the springs and soak in one of the runoff channels from the scalding thermal pools, but for much of the year the hot springs bubbled in solitude. In the early 1990s Mickey Hot Springs caught the attention of the press when the geyser began to erupt regularly to a height of 8 feet or more. More than 2,000 people visited the hot springs that year to see the eruptions.

Concerned with the impact of increasing numbers of visitors, the BLM conducted an environmental assessment on the hot springs in 1995. Two years later the BLM built a fence around the hot springs to keep both cattle and cars from damaging the fragile thermal features. (According to a BLM spokesman, a few calves from the Alvord Ranch that graze on BLM land have fallen into the hot springs over the years.)

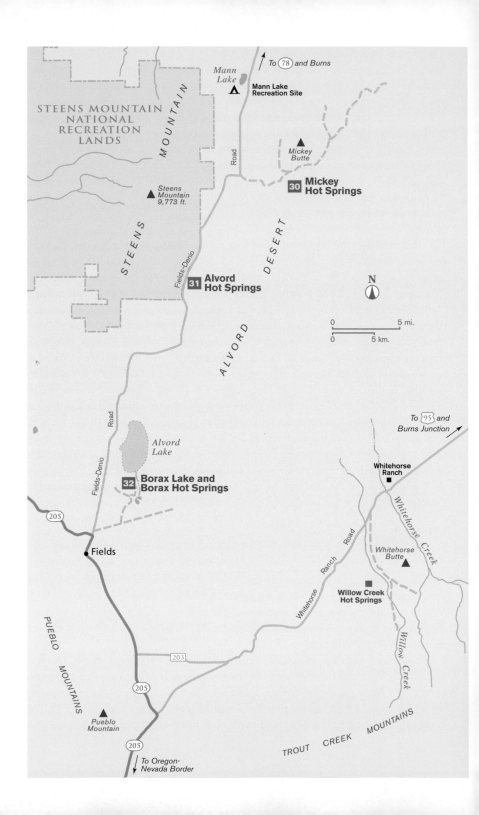

To 78 and Burns

Mann Lake

Mann Lake
Recreation Site

STEENS MOUNTAIN
NATIONAL
RECREATION
LANDS

MOUNTAIN

Road

Mickey
Butte

30 Mickey
Hot Springs

Steens
Mountain
9,773 ft.

STEENS

DESERT

Fields-Denio

31 Alvord
Hot Springs

N

0   5 mi.

0   5 km.

ALVORD

Road

Fields-Denio

Alvord
Lake

To 95 and
Burns Junction

Whitehorse
Ranch

32 Borax Lake and
Borax Hot Springs

Whitehorse Creek

205

Whitehorse Ranch Road

Whitehorse
Butte

Fields

Willow Creek
Hot Springs

Willow Creek

PUEBLO

203

205

MOUNTAINS

Pueblo
Mountain

TROUT CREEK MOUNTAINS

205

To Oregon-
Nevada Border

Heed the warning signs when touring Mickey Hot Springs.

A zigzag opening in the fence now allows visitors to pass through while keeping cattle outside.

A sad chapter occurred in the history of Mickey Hot Springs in the autumn of 1996. A hunter found the dead body of a man in a 117-degree-F thermal pool. A total eclipse of the moon had been predicted the night before, and police surmised that the man had watched the lunar event while soaking in the solitude of the desert springs. It's unknown how long the man soaked in the hot pool or whether the high water temperature and long soaking time led to a heart attack.

31. ALVORD HOT SPRINGS

See map on page 148.

General description: A pair of rustic soaking pools on the edge of the Alvord Desert in southeastern Oregon.

Location: Southeast Oregon, 107 miles southeast of Burns in the Alvord Desert.

Development: A small open-air bathhouse containing 2 soaking pools was built near the hot springs decades ago. A convenience store, caretaker's trailer, and restroom building were added in 2013. Although not a totally undeveloped soak, the isolated, remote location will still bring smiles to the faces of all but the most fanatic hot springs purists.

Best time to visit: Autumn, winter, and spring are excellent times to soak at Alvord Hot Springs. Visiting during the desert heat of July and Aug is less appealing; it's best to leave the desert (and the hot springs) to the lizards, and wait until cooler weather.

Restrictions: A soaking fee is charged for each bather. Nudity is the norm in the bathing pools.

Access: The hot springs are privately owned by the Alvord Ranch (which formed a limited liability corporation to manage the hot springs). The landowner graciously allows the public to use the hot springs for a nominal fee. Any vehicle can make the trip on the paved and graveled Fields-Denio Road to the parking area adjacent to the hot springs. A rare snowstorm may block portions of Fields-Denio Road in the winter.

Water temperature: The hot springs emerge along a series of 170-degree-F seeps on the east side of Fields-Denio Road. Hot water then meanders 100 yards east, cooling to about 120 degrees F by the time it reaches the 2 soaking pools. The actual soaking temperature usually varies between 100 and 115 degrees F, depending on how rapidly the hot springs water is piped into the pools.

Nearby attractions: The 30-mile-long Steens Mountain towers more than 5,000 feet above the Alvord Desert. Steens Mountain Loop Road, a scenic 52-mile traverse from the lower foothills to the mountain summit, starts 8 miles south of Frenchglen (northwest of Fields). The massive ridge of Steens Mountain to the west creates a rain shadow farther east, creating the harsh, dry environment of the Alvord Desert. East of Alvord Hot Springs stretch the usually dry desert lake beds (or "playas"). During occasional wet years the lake beds may contain a shallow layer of water for a couple of months, but for most of the year, the playas are dry and dusty. Plan on spending a couple of days exploring other geothermal springs in the Alvord Desert, including Borax Lake, Mickey Hot Springs, and Willow Creek Hot Springs.

Services: A small convenience store, restrooms, and changing rooms are located a few hundred yards from the soaking pool. About a dozen picnic tables and fire pits are available. The on-site caretaker can sell you water, ice, and bath towels, in case you've forgotten any of those items. Six bunkhouses for overnight guests are available (but bring your own sleeping bags or other bedding).

The tiny town of Fields (population 14) is located 23 miles to the south, where food, gas, drinking water, and lodging are available. The small motel in Fields has 2 clean but simple rooms available (each room has 2 bedrooms with queen beds and a bathroom). It's best to call ahead for motel reservations (541-495-2275). The Fields Café serves huge breakfasts, hamburger lunches, and dinner. The cafe is famous for its thick milkshakes—check the chart on the restaurant wall to see how many shakes have been sold since the beginning of the calendar year.

Camping: Tent camping is allowed across the road from the hot springs (the camping fee includes soaks at the hot springs). Close to a dozen spots are available and provide a gorgeous view to the east over the Alvord Desert. Many campers have commented on the marvelous sunrises and sunsets that can be seen from the campground. Camping is also available at the BLM's Mann Lake Recreation Site, 18 miles north of Alvord Hot Springs on Fields-Denio Road. Drinking water is available at the small store at the hot springs, so stock up before you venture deeper into the eastern Oregon desert. Four BLM campgrounds are also available on Steens Mountain Loop Road on the west side of Steens Mountain.

Maps: BLM Steens High Desert Country map; *DeLorme: Oregon Atlas & Gazetteer,* page 86, C3.

GPS coordinates: N42.5439' / W118.5331'

Contact info: Alvord Hot Springs LLC, 36095 E Steens Rd, Princeton, OR 97721; www.alvordhotsprings.com

Finding the springs: From Burns drive 65 miles southeast on OR 78. Turn south onto Fields-Denio Road (also called Fields-Follyfarm Road) and drive 42 miles toward Fields (this road is partially paved and partially gravel, so be prepared for a slow and bumpy trip). If you've been visiting Mickey Hot Springs to the north, Alvord Hot Springs is 10.5 miles farther south on Fields-Denio Road from the intersection with the turnoff to Mickey Hot Springs. You'll see the steaming hot springs and tin-sided bathhouse about 100 yards east of the road, and a campground on the west side of the road. Turn east and park next to the small convenience store. The caretaker will likely be in the store or in his nearby trailer—you'll need to hand over the fee for a soaking pass for each of your passengers. It's best to make overnight reservations for the bunkhouses or camping sites ahead of time—they can fill up quickly on busy weekends or during the summer.

THE HOT SPRINGS

The pools at Alvord Hot Springs have hosted decades of local ranchers, bird and elk hunters, and tourists enjoying Steens Mountain and the Alvord Desert. The two adjoining soaking pools are each about 8 feet square and 4 feet deep. Corrugated tin sheeting surrounds the northernmost pool. Judging from the numerous bullet holes visible in the metal walls, this pool has seen more than a few wild parties. There's no roof over either pool, which allows great views of the stars at night. A small bench sits on the wooden deck on the east side of the enclosed pool. South of the enclosed pool is the open-air pool, which tends to be the more popular of the two because of its unobstructed views of Steens Mountain and the dry desert lake beds. Tubs from

Old washing machine tubs provide comfortable soaking seats at Alvord Hot Springs.

the interiors of old washing machines have been placed upside down in the pools to provide quirky but comfortable seating for bathers.

Alvord Hot Springs has a unique temperature-control system. Two-inch steel pipes divert the 120-degree-F water from the hot creek to the lip of the soaking pools. The pipes quickly fill the soaking pools to an uncomfortably hot temperature if the hot-water flow is left unchecked. Fortunately the ends of the hot-water pipes can be raised about a foot into the air and placed on a concrete block, which allows gravity to stop the flow of hot water. Once the hot-water flow is stopped, the pools will eventually cool to a comfortable bathing temperature.

There's also a wooden stopper in the bottom of each pool, which can be removed to allow the water to drain completely before filling for a fresh soak.

You never know what the water temperature will be in the Alvord Hot Springs pools when you first arrive, as it depends on whether the previous bathers had left the hot water flowing into the pool or diverted the flow to cool things down. Either way it usually only takes a half hour or so to adjust the hot-water flow to a comfortable soaking temperature.

Alvord Hot Springs in the shadow of Steens Mountain

The runoff from Alvord Hot Springs provides moisture for one of the few bird habitats in the Alvord Desert. Long-billed curlews nest in this area, and American avocets, snowy plovers, and killdeers can be found farther east on the playa.

History

The Davis family has owned the nearby Alvord Ranch and Alvord Hot Springs since the 1970s. For over forty years the family allowed public soaking at the hot springs without any oversight. Due to the increasing popularity of the hot springs, the family decided in the fall of 2013 to house a caretaker on-site and charge a nominal per person soaking fee to help defray the costs of keeping the hot springs site clean and safe.

32. **BORAX LAKE AND BORAX HOT SPRINGS**

See map on page 148.

General description: A string of more than a dozen hot springs that march north along a fault line in the Oregon desert. The thermal water from the southernmost hot springs creates the 5-acre warm-water Borax Lake.

Location: Southeast Oregon, 7.5 miles north of Fields in the Alvord Desert.

Development: Undeveloped. A century ago the mineral-rich borax found in the soil surrounding Borax Lake was concentrated in steel vats and shipped to California. The ruins of these vats can still be seen near the edge of Borax Lake, but nothing else remains of this commercial development.

Best time to visit: Driving on the dirt roads in the area could be difficult after a spring rainstorm, so watch the weather forecast. The desert heat in July and Aug can be brutal—a Sept or Oct visit may be the most pleasant time of year for a visit.

Restrictions: The Nature Conservancy doesn't permit soaking in Borax Lake or the nearby hot springs, both to protect the fragile nature of the resource as well as to protect the threatened fish species that is found only in Borax Lake. Use caution around the unmarked hot springs, some of which approach 180 degrees F. Signs posted by the BLM warn visitors that they could break through the thin crust surrounding some of these hot springs, so avoid standing too close to the edge of the hot pools. Also, the BLM has measured arsenic levels in the lake and hot springs more than 25 times higher than acceptable drinking standards, so it's best not to drink the water. No dogs are allowed on the property, so leave your puppy behind.

Access: High-clearance vehicles are recommended on the bumpy dirt access road, but if you take your time, almost any vehicle can make the trip. The roads can turn to impassable muck after a rainstorm.

Water temperature: The 5-acre Borax Lake ranges from 84 degrees F near its shallow shore to 95 degrees F in the deep center of the lake. The source of the lake's warmth is a hot spring that is perhaps hundreds of feet deep at its center. This geothermal resource may pump more than 1,000 gallons per minute of 200-degree-F water into the lake bottom. The series of hot springs north of Borax Lake vary in temperature from 90 degrees F to more than 170 degrees F.

Nearby attractions: The 30-mile-long Steens Mountain towers more than 5,000 feet above the Alvord Desert. Steens Mountain Loop Road, a scenic 52-mile traverse from the lower foothills to the mountain summit, starts 8 miles south of Frenchglen (northwest of Fields). The massive ridge of Steens Mountain to the west creates a rain shadow farther east, creating the harsh, dry environment of the Alvord Desert. North of Borax Lake stretches desert lake beds (or "playas") that seem as smooth and flat as a pool table. During occasional wet years the lake beds may contain a shallow layer of water for a couple of months, but for most of the year the playas are dry and dusty. Plan on spending a couple of days exploring other geothermal springs in the Alvord Desert, including Alvord Hot Springs, Mickey Hot

Springs, Willow Creek Hot Springs, and Bog Hot Springs just across the border in northern Nevada.

The fossil beds located in the Trout Creek area southeast of Fields are worth an afternoon visit. If you're lucky, you may find a beautiful jet-black fossil of prehistoric fern, which stands in stark contrast to the surrounding snow-white rock strata. Stop at the cafe in Fields for directions to the fossil beds.

If you have more time, consider hiking one of the rarely visited gems of the Oregon State Parks trail system—the 100-mile-long Desert Hiking Trail. The trail starts at the Nevada border south of Fields and heads northwest to the top of Steens Mountain.

Services: No services are available at Borax Lake or Borax Hot Springs. The tiny town of Fields (population 14) is located 7.5 miles to the south, where food, gas, drinking water, and lodging are available. The small motel in Fields has 2 clean but simple rooms available (each room has 2 bedrooms with queen beds and a bathroom). It's best to call ahead for motel reservations (541-495-2275) or check their Facebook page (www.facebook.com/thefieldsstation/). The Fields Café serves huge breakfasts, hamburger lunches, and dinner. The cafe is famous for its thick milkshakes—check the chart on the restaurant wall to see how many shakes have been sold since the beginning of the calendar year.

Camping: Camping isn't allowed at the lake or the hot springs, but plenty of options abound on nearby BLM land, where camping permits are not required. Take one of the side roads that wander into the foothills of Steens Mountain northwest of Borax

Lake and camp anywhere on BLM land. (Check the BLM map to ensure you are on public property before rolling out your sleeping bag for the night.) Camping is also available at the BLM's Mann Lake Recreation Site, 31 miles north of Fields on Fields-Denio Road. (No drinking water is available at Mann Lake, so be sure you have plenty with you.) Four BLM campgrounds are available on Steens Mountain Loop Road on the west side of Steens Mountain. Camping spots are also available for a fee at Alvord Hot Springs.

Maps: BLM Steens High Desert Country map; *DeLorme: Oregon Atlas & Gazetteer,* page 86, D3.

GPS coordinates: N42.3275' / W118.6048'

Contact info: The Nature Conservancy of Oregon, 821 SE 14th Ave., Portland, OR 97214; (503) 802-8100; www.nature.org/en-us/get-involved/how-to-help/places-we-protect/borax-lake; oregon@tnc.org.

Finding the springs: From the town of Fields, drive north 1.2 miles to the intersection of OR 205 and Fields-Denio Road. Proceed 0.5 mile north on Fields-Denio Road and then turn east onto a dirt road next to an electrical substation. Head northeast on this dirt road for 2.1 miles, parallel to the overhead power lines. Turn north onto a washboarded road for 2 miles. You'll reach a closed gate with a BLM sign warning about the danger of the scalding water at the hot springs. Park your car at the gate and walk about 1 mile on a dirt road that takes you past Lower Borax Lake. Continue past Lower Borax Lake for another 0.5 mile until you see the rusting ruins of the steel vat from the borax-processing operation of a century ago. Borax

Lake is less than 100 yards east of these ruins.

To reach the series of thermal pools collectively named Borax Hot Springs, walk about 0.1 mile on the dirt road that leads northwest from Borax Lake. You'll start to see a string of hot springs about 10 feet east of the road that stretch for about a mile along a fault line running north across the desert. Enjoy your walk along the road and path that border the bubbling desert springs. When you run out of springs, you'll need to backtrack south on this road past Borax Lake to your car.

THE HOT SPRINGS

Borax Lake and the nearby string of hot springs are one of the most unusual geothermal areas in Oregon. The hot pools seem to march single file along an imaginary line that follows a geologic fault beneath the desert's surface.

The 5-acre Borax Lake looks nondescript from the shore, but an aerial photo tells a much more interesting geologic story. The lake is actually the outflow from a hot spring located deep beneath the center of the lake. Borax Lake is only a few feet

Borax Lake

deep until the center is reached, when the lake bed disappears over the dark core of a 95-degree-F hot spring that's perhaps hundreds of feet deep. Unfortunately swimming is no longer allowed in the lake (or nearby hot springs), so you'll have to enjoy the view from the shoreline.

After you've explored Borax Lake and the nearby ruins of the old borax works, walk along the road north of the lake to see the row of steaming hot springs. Many of these pools approach 170 degrees F or more, so stay away from the fragile pool edges as you head north on the road. The northernmost pool is located about 0.6 mile from Borax Lake, at the end of the string of hot springs where the road turns west away from the pools. When you've finished exploring these hot springs, backtrack to Borax Lake.

History

In 1897 Charles Taylor heard rumors of extensive deposits of alkali in the southern Alvord Desert. Taylor operated a small borax mine in Nevada and was interested in finding other borax deposits that he could develop. Taylor traveled across the Nevada-Oregon border and discovered large deposits of snow-white sodium borate surrounding Borax Lake. A rancher named Robert Doan owned the mineral-encrusted

One of the many hot pools that dot the ancient lake bed at Borax Hot Springs

A sinter deposit arches through one of the pools at Borax Hot Springs.

property, which he considered useless for ranching. It didn't take much convincing for Doan to accept Taylor's offer of $7,000 for the 3,000 acres of "worthless" land surrounding Borax Lake.

By 1898 Taylor and his partner, John M. Fulton, had moved their borax operation from Nevada to Borax Lake. Fulton hired a chemist named Christian Ollgard to devise a unique method for processing the raw sodium borate into the refined crystalline form needed by customers on the West Coast. Key to the processing operation were two steel boiling vats that Ollgard built next to Borax Lake. The round-bottomed vats could each hold 6,000 gallons of borate solution. Chinese laborers were hired to fill the vats with raw sodium borate gathered from the desert surface. Ollgard then piped 97-degree-F water from the warm center of Borax Lake into the vats to cover the borate minerals. Sulfuric acid was added to the solution as a precipitating agent. Sagebrush was then gathered and burned under the vats for several hours until the solution was heated to boiling. The solution was then allowed to cool, which caused the raw borate to form pure crystalline borax. The snow-white borax crystals were bagged in 90-pound sacks and hauled in wagons by up to two dozen mules 150 miles south to Winnemucca, Nevada. More than 55,000 pounds of borax were hauled to Nevada every week.

Taylor and Fulton named their borax operation the "Twenty Mule Team Borax Company," after the hauling method they devised to carry the borax to Nevada. Unfortunately they failed to legally register this name, and within a year a competing borax company in California had taken claim to the now-legendary brand. Taylor and

An old steel vat used to boil and concentrate borax near Borax Lake

Fulton were forced to change their company name to "Rose Valley Borax Company," a prettier but less memorable moniker.

Taylor and Fulton sold their borax operation to Christian Ollgard in 1902, but by 1907 the declining quality of borate deposits from the surrounding desert led to the closure of the facility. At present only the rusting steel boiling vat and the ruins of a nearby sod house remind visitors of the bustling borax operation and the twenty-mule-team wagons that once passed through the valley.

For several decades after the Rose Valley Borax Company ceased operation, Borax Lake and the nearby hot springs received little attention and few visitors. All that changed in 1980, when an endangered fish was discovered living in the lake's warm, brackish waters. The new species of fish was named the Borax Lake chub (*Gila boraxobius*). Only a few thousand of this half-inch-long fish survive on the edge of Borax Lake, where the temperature remains a constant 85 degrees F. Shortly after the chub's discovery in 1980, the US Fish and Wildlife Service declared the Borax Lake chub an endangered species. The species has not been found anywhere else in the world.

The geothermal-energy potential of the area around Borax Lake has also attracted attention. In 1990 the BLM gave permission to a California energy company to drill a geothermal test well near Borax Lake. The energy company hoped to find super-heated water that could be used in a turbine to generate electricity. Environmentalists expressed concern that drilling near Borax Lake might disrupt the natural hot springs feeding the lake and change the delicate aquatic environment needed to support the Borax Lake chub. In response to the environmental threats to the endangered fish, the Nature Conservancy obtained a lease to manage Borax Lake and 320 acres of adjoining public land. The stewardship of the Nature Conservancy, the Bureau of Land Management, and the Oregon Department of Fish and Wildlife has helped increase the numbers of this fish in the lake over the past thirty-five years, and its status may soon be upgraded from "endangered" to "threatened."

33. BOG HOT SPRINGS (NORTHWEST NEVADA)

General description: A high desert hot springs flows into an irrigation ditch that contains several warm soaking pools.

Location: Northwest Nevada, just south of the Oregon border, about 130 miles east of Lakeview, Oregon.

Development: Undeveloped except for some man-made rock dams across the irrigation ditch.

Best time to visit: Open year-round. The soaking pools may be a bit crowded on Memorial Day and other holiday weekends, as well as during fall hunting season. Consider a visit during a full moon—a desert soak under a bright lunar sky is unforgettable.

Restrictions: The hot springs and the irrigation ditch and pools are privately owned, but the owners allow free unlimited soaking. Expect nudity.

Access: Any vehicle can make the trip along NV 140. The last 4 miles of graded dirt road to the hot springs may be muddy during wet weather—keep an eye on the skies and the weather forecast.

Water temperature: The hot springs emerge from the ground at about 130 degrees F but cool as they flow through the irrigation ditch. Bathers can find a range of temperatures along the ditch to suit their soaking preferences. The largest soaking pool is about 100 degrees F.

Nearby attractions: Sheldon National Wildlife Refuge (Nevada), Steens Mountain, and the Alvord Desert (Oregon).

Services: No services at the hot springs. Denio Junction, about 13 miles east of the hot springs, has a 7-room motel, RV spaces, restaurant, small grocery store, ice cream, and gas (call 775-941-0171 for motel reservations). Fill up your tank and water supply here before heading on to Bog Hot Springs. The tiny town of Fields, Oregon, is located 38 miles north, where food, gas, drinking water, and lodging are available. The small motel in Fields has 2 clean but simple rooms available (each room has 2 bedrooms with queen beds and a bathroom). It's best to call ahead for motel reservations (541-495-2275). The Fields Café serves huge breakfasts, hamburger lunches, great milkshakes, and dinner.

Camping: Camping is allowed pretty much anywhere around the hot springs and irrigation ditch (you may be next to many other car campers if you stay near the springs, especially on holiday weekends). Consider camping farther away from the springs on any BLM land for more privacy.

Maps: Nevada State Highway Map; *DeLorme: Nevada Atlas & Gazetteer,* page 19, B9; USGS Denio, NV (1:100,000).

GPS coordinates: N41.9239' / W118.8036'

Contact info: None available.

Finding the springs: If you're coming from Oregon after a tour of the Alvord Desert hot springs, drive south on OR 205 for 24 miles to Denio Junction (on the Oregon-Nevada border). Stop for water, gas, and food at the store in Denio Junction if you need to stock up. Take NV 140 west from Denio Junction for 9.1 miles, then turn north off the highway onto a gravel road (you'll see a Bog Hot Springs sign at the turnoff). At about 3.5 miles you'll pass Bog Ranch and a large reservoir. Continue driving about 0.7 mile farther north to the steaming irrigation ditch and parking area that marks Bog Hot Springs. Park your car, grab your towel, and slip into the hot pools just off the parking area. Work your way up the ditch for warmer soaks if you desire.

THE HOT SPRINGS

Although it's officially a Nevada hot springs, many Oregonians claim this as one of their own, since the main highway between Fields and Lakeview in Oregon dips down into Nevada and passes near this jewel of a soak. Since it's only about 5 miles off NV 140, Bog Hot Springs is a great hot springs visit if you've been soaking in the Alvord Desert area north of Fields and are on your way to south-central Oregon to visit more thermal areas.

The privately owned hot springs are a steamy 131 degrees F from the ground, but the water then flows for a few hundred yards to the Bog Reservoir, where it is used for irrigation by a ranch. Between the reservoir and the hot springs is a large irrigation ditch, which is ground zero for Bog soaks. Past visitors have built small rock dams in

Bog Hot Springs near the Oregon-Nevada border CALEB MORRIS

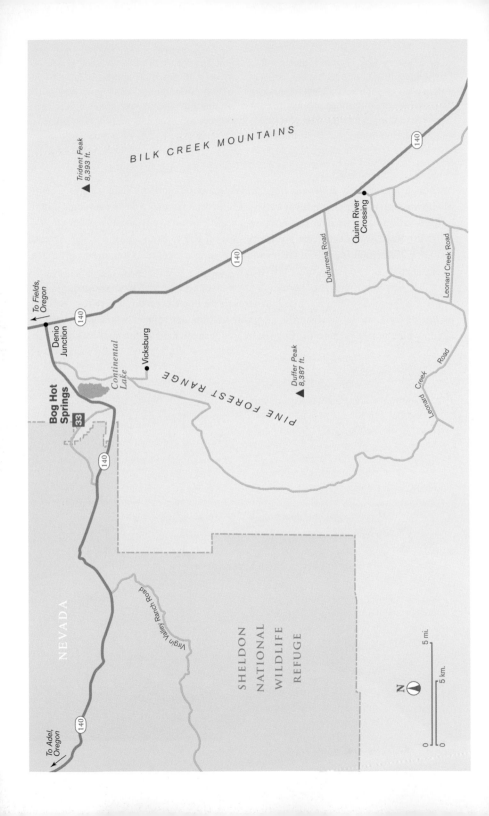

BILK CREEK MOUNTAINS

Trident Peak
8,393 ft.

To Fields,
Oregon

Denio
Junction

Continental
Lake

Vicksburg

Bog Hot
Springs

33

140

140

140

140

140

140

Quinn River
Crossing

Dufurena Road

Leonard Creek Road

Leonard Creek Road

Duffer Peak
8,387 ft.

PINE FOREST RANGE

NEVADA

Virgin Valley Ranch Road

SHELDON
NATIONAL
WILDLIFE
REFUGE

To Adel,
Oregon

N

5 mi.

5 km.

0

0

Soaking pools are scattered throughout the irrigation ditch at Bog Hot Springs. CALEB MORRIS

the ditch to create soaking pools. You'll likely see other cars pulled up near the ditch at the prime soaking spots, but there is usually plenty of room for all to enjoy the best soaking areas. There are many spots along the ditch worth dipping a toe in, so move upstream from the lower pools if you want more privacy.

34. WILLOW CREEK (WHITEHORSE RANCH) HOT SPRINGS

General description: One of Oregon's most remote rustic soaking pools, surrounded by sagebrush and low mountain ridges.

Location: Southeast Oregon, 138 miles southeast of Burns.

Development: The hot springs are part of a BLM campground and recreation area, but development is limited to a few fire pits and a well-built privy. The area surrounding the natural hot springs has been enlarged and deepened, and a small concrete-and-rock dam separates the smaller, hot soaking pool from the larger, cooler pool.

Best time to visit: May to Oct are prime times for visiting, although the hot springs are open year-round. You may want to avoid the searing hot weather in midsummer, as well as times when the roads turn to muck due to heavy rains. The campground is often filled with hunters in late fall during hunting season.

Restrictions: Camping isn't allowed within 100 feet of the hot springs. The BLM has imposed a 14-day limit to camping at the hot springs and elsewhere on land in the BLM Burns District. Nudity is common in the hot springs pools.

Access: Just about any vehicle can make the trip during dry weather. The last couple of miles to the hot springs can become impassable during rainy weather, when the normally dry and dusty road is transformed into deep, slippery mud.

Water temperature: The soaking pool containing the main hot springs averages 102 degrees F. The cooler overflow pool varies between 85 and 95 degrees F.

Nearby attractions: Willow Creek Hot Springs is sandwiched between the Trout Creek and Oregon Canyon Mountains, and 7 wilderness-study areas lie within 15 miles of Willow Creek. Backpacking along the low mountain ridges and vegetated canyons is a popular activity. Hunters use the hot springs as a staging area for mule deer and sage-grouse hunts. Fishing is allowed in some area streams, although Willow Creek, which flows by the hot springs, is off-limits to anglers. A rare inland species of cutthroat trout—the Lahontan cutthroat—is the only fish found in Willow Creek and is listed by the federal government as a threatened species. Take a walk along Willow Creek and see if you can spot this rare fish, but leave your fishing pole in the car.

Services: None at the hot springs. Because it's more than an hour's drive to the nearest store, be sure to have a full tank of gas and plenty of food and water before starting your trip.

Camping: Willow Creek Hot Springs is a BLM recreation site; camping areas with fire pits are located within a few hundred yards of the hot springs. Camping spots are first-come, first-served, and camping permits are not required. If others are camped near the fire pits and you desire more privacy (and a more natural setting), take your sleeping gear a few hundred yards south and find a secluded spot near the shrub-lined banks of aptly named Willow Creek. Dispersed camping is also allowed in the

Willow Creek Hot Springs

surrounding wilderness-study areas and other nearby BLM lands; permits are not required.

Maps: BLM Steens High Desert Country map; *DeLorme: Oregon Atlas & Gazetteer,* page 86, D4.

GPS coordinates: N42.2750' / W118.2656'

Contact info: Bureau of Land Management, Burns District Office, 28910 Hwy. 20 W., Hines, OR 97738; (541) 573-4400; www.blm.gov/office/burns-district-office.

Finding the springs: Visitors to Willow Creek Hot Springs often first loop south through the Alvord Desert for a day or two, visiting Alvord Hot Springs, Mickey Hot Springs, and Borax Lake. If you are already in the

Alvord Desert, start your drive to Willow Creek Hot Springs at the tiny town of Fields (stock up at the local store with drinking water and gas). From Fields drive south on OR 205 toward the Nevada border for 8.2 miles to CR 203 (Whitehorse Ranch Road). Drive east on CR 203 for 25 miles (the road turns northeast about 10 miles into this drive), then look for a poorly marked dirt road on the right (south). Drive a total of 2.3 miles south on the dirt road until you see a rocky rise on the right with a concrete outhouse at its base. Take the right fork onto the dirt road toward the outhouse, passing the hot springs pool as you enter the campground.

The trickiest part of this drive is finding the turnoff to the dirt road from CR 203. Locals use the White-horse Ranch headquarters as their main point of reference, so if you overshoot the dirt road, keep going another 2 or 3 miles on CR 203 to the main gate of the Whitehorse Ranch. Turn around and reset your odometer, then drive 2.7 miles back southwest on CR 203 and look for the dirt road to the south that leads to Willow Creek Hot Springs.

You can also drive south to Wil-low Creek Hot Springs from Burns Junction, about 91 miles southeast of Burns on OR 78. From Burns Junction head south on US 95 for 21 miles to the Whitehorse Ranch Road intersec-tion (CR 203). Drive southwest for 21 miles on CR 203 to the Whitehorse Ranch headquarters. Continue south-west past the ranch headquarters for 2.7 miles to the dirt road on the left (south).

THE HOT SPRINGS

Willow Creek Hot Springs is perfect if you're seeking a truly isolated soak far from population centers. The hot springs pool measures about 10 feet by 50 feet and is divided in half by a low concrete dam. The smaller soaking pool averages 102 to 104

Camping and parking are just a few feet from Willow Creek Hot Springs.

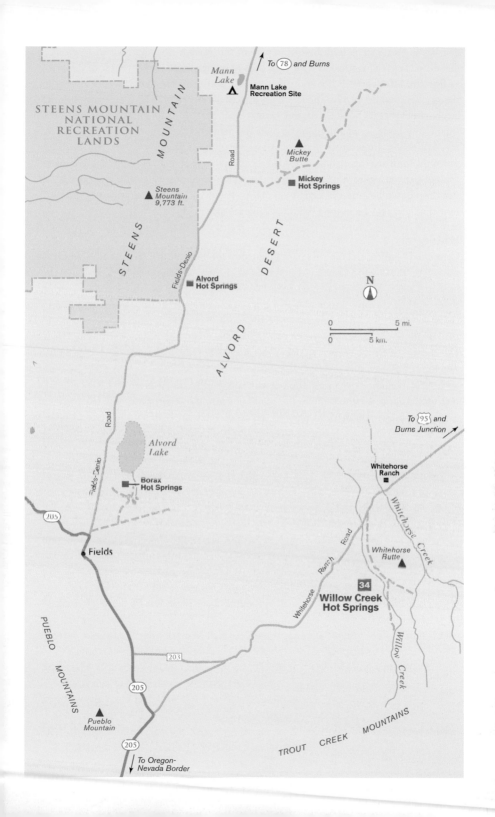

degrees F. Stone steps lead down the side of the smaller hot pool, which is about 3 feet deep. Try feeling around with your toes in the mud and gravel center of this pool to find the 110-degree-F hot spots that are the source of the hot water. Water from the smaller pool flows over the low dam into the larger pool. The temperature in the larger pool averages between 85 and 95 degrees F. Start your soaking experience in the smaller but hotter pool and then slip over the dam into the larger pool to cool down.

It's almost imperative that you stay overnight at Willow Creek Hot Springs, given the long driving distances needed to reach this spot. Plan on arriving by midafternoon and setting up your camp, and then soak in the hot springs while watching the sun drop down beyond the sagebrush plains. If you can rouse yourself early the next morning, you may be in for a fabulous sunrise soak.

History

Camp C. F. Smith was the first significant development in the area, built as an army post along the Oregon Central Military Road in 1866. When the army abandoned the post in 1869, a Virginian named John Devine acquired the buildings and converted them and the surrounding land into a major ranching operation. Devine was said to be a dashing figure who often surveyed his ranch properties from the back of a beautiful white horse, which led to the naming of the Whitehorse Ranch (Willow Creek Hot Springs is better known to many locals as Whitehorse Hot Springs). Besides his ranching interests, Devine was devoted to raising racehorses, and in 1879 he built an impressive horse barn near the present entrance to the Whitehorse Ranch headquarters. A large weathervane in the shape of a white horse sits atop the barn, which serves as a major landmark in the isolated ranching country of southeastern Oregon. The Whitehorse Ranch is one of Oregon's largest ranches, encompassing more than 68,000 acres.

35. SNIVELY HOT SPRINGS

General description: A popular riverside soak that can be washed out by high runoff from upstream dam releases and spring snowmelt.

Location: Southeast Oregon, 30 miles south of the town of Ontario, near the Oregon-Idaho border in the scenic Owyhee River canyon.

Development: None.

Best time to visit: Since the hot springs can be submerged in the Owyhee River at high water levels, it's best to visit in late fall, winter, and early spring before irrigation season starts, since releases from the Owyhee Dam will raise river levels. Heavy snow in the winter can also raise water levels during spring runoff.

Restrictions: Snively used to be open without restrictions, but in 1998 the BLM closed the recreational site from sunset to sunrise due to complaints of fighting and alcohol abuse. No camping is allowed at the site, and no glass containers are allowed within 30 feet of the hot springs due to concerns over broken glass in the water and on the trails.

Access: Any vehicle can make the trip. The access road passes right next to the hot springs on the riverbank.

Water temperature: The hot springs emerge from a cement cistern above the riverbank at around 135 degrees F and then form a hot-water stream that flows about 100 yards before it enters the river. The hot water mixes with river water to form 2 or 3 soaking pools. The pool temperatures vary widely, depending on their distance from the river (and the amount of cold river water that mixes in each soaking pool). Temperatures in the pools

closest to the riverbank are usually the hottest and may hit 120 degrees F. Temperatures farther out in the river can vary from 102 to 110 degrees. During periods of high water in the river, the entire hot springs area may be washed out, and no soaks will be found.

Nearby attractions: Lake Owyhee State Park, which borders the 53-mile-long Owyhee Reservoir, is an excellent base for boating, fishing, and hiking.

Services: None at the hot springs, other than an enclosed pit toilet.

Camping: There is camping available in nearby Lake Owyhee State Park. Camping is also allowed on any nearby BLM land (look for primitive campsites near the riverbanks).

Map: *DeLorme: Oregon Atlas & Gazetteer,* page 83, B10.

GPS coordinates: N43.7300' / W117.2034'

Contact info: Bureau of Land Management, Vale District Office, 100 Oregon St., Vale, OR 97918; (541) 473-3144, www.blm.gov/office/vale-district-office.

Finding the springs: From Ontario drive south on OR 201 for 20 miles to Owyhee Junction. Continue south for about 8 miles on the road toward Lake Owyhee State Park (follow the signs). Watch for a pipeline that spans the river canyon and drive an additional 1.5 miles. You'll see a BLM sign on the side of the road for Snively Hot Springs. Park at the large Snively parking lot and walk to the riverbank. Depending on the river water level, you'll see the hot springs pools in the river extending out from the bank.

Snively Hot Springs on the banks of the Owyhee River WAYNE ESTES

(In high-water season you may only see the river flowing by, since the hot springs are often submerged in the spring runoff.)

THE HOT SPRINGS

Although it's a bit of a drive from most anywhere, Snively Hot Springs offers some wonderful soaking opportunities, as long as the river level isn't so high that the soaking pools are washed out. If you visit during a low-water time of year, you'll find two or three pools (volunteers rebuild the pools with river rocks every year). Pool depth can vary from quite shallow to 3 feet or more. Be cautious when entering the pools—the geothermal water that enters the river can be scalding, and unless it mixes with sufficient cool river water, the soak will be too hot for comfort. You may need to stir the hot water that floats on top of the cooler river water to find the right consistent water temperature for a soak. A little more than 1 mile before Snively Hot Springs, you will see the huge pipeline transferring waters of Deer Butte Hot Springs from the source in the mountains to the small pond located on the other side of the road. The springs are readily accessible throughout the year, although the pools may submerge during the spring runoff, leading to temporary closure, particularly in the high trout fishing season. Just before and after the spring, Snively provides a matrix of lush colors from the horizon, extremely comfortable temperatures, and the main course of a hot mineral-water therapy.

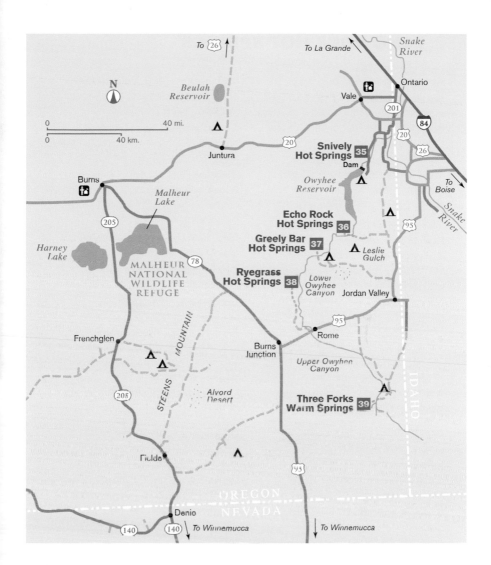

The extremely hot water of the thermal springs emerges out of the ground 300 feet from the river. The hot water flows down through a ditch to the manually built pool and mixes with the cold river water. The temperatures may vary from cold to very hot depending on your location in the pool. You can find the perfect temperature for soaking by moving toward or farther away from the source. The pool with the gravel bottom is approximately 20 feet in diameter and 2 to 3 feet deep.

Due to the overuse and abuse of the recreation area, the BLM has closed the area from dusk until dawn. Also, glass containers are not permitted within 30 feet of the springs, and camping and fires are not allowed.

36. ECHO ROCK HOT SPRINGS

See map on page 173.

General description: Two small soaking pools on the banks of the Owyhee River, just upstream from impressive towers of volcanic tuff.

Location: Southeast Oregon, in the Owyhee River canyon, 4 miles upstream from the Leslie Gulch boat ramp.

Development: None.

Best time to visit: Visiting in the spring and early summer is best if you'd like to raft upstream on the Owyhee River to the hot springs. When the water level drops in the fall, it's possible to hike along the shallow river the 4 miles from the Leslie Gulch boat ramp upstream to the hot springs.

Restrictions: None.

Access: Any vehicle can make the trip to the Leslie Gulch boat ramp (although parts of the road can be impassable when muddy during a rainstorm). Accessing the hot springs from Leslie Gulch requires either a 4-mile hike or a raft trip up the Owyhee River.

Water temperature: The upper pool is about 106 degrees F, and the lower pool is around 103 degrees F.

Nearby attractions: Hiking or rafting to Echo Rock Hot Springs will take you through an amazing collection of volcanic ash tuff formations that tower above the canyon walls. You may also see California bighorn sheep, part of a herd of 200 that were introduced into

the canyon in 1965. Mule deer, elk, and a herd of wild horses are found in the area too.

Services: None. Bring plenty of drinking water or a water-purification system.

Camping: Slocum Creek Campground is located near the Leslie Gulch boat ramp. Camping is limited to 14 days. No drinking water is available.

Map: BLM Owyhee Canyon Country Vale District South map.

GPS coordinates: N43.3017' / W117.3830'

Contact info: Bureau of Land Management, Vale District Office, 100 Oregon St., Vale, OR 97918; (541) 473-3144; www.blm.gov/office/vale-district-office.

Finding the springs: From Boise, Idaho, take US 95 west for 56 miles to McBride Creek Road (also called the Leslie Gulch Road junction). Take McBride Creek Road west for 8 miles to Rockville and proceed north for a mile to Leslie Gulch Road. Follow Leslie Gulch Road for about 15 miles to the Leslie Gulch boat ramp in the Owyhee River canyon. From the boat ramp it's a 4-mile hike on the south side of the river or a paddle upstream to Echo Rock Hot Springs. The hot springs are located directly across from a tall rock wall (called Echo Rock). Look for a white pipe across the river from the rock wall—the pipe marks the location of Echo Rock Hot Springs.

THE HOT SPRINGS

The last 4-mile journey to reach Echo Rock Hot Springs is as enjoyable as the soak—the towers of volcanic tuff on the side of the Owyhee River canyon look like something right out of the *Lord of the Rings* trilogy. The hot springs consist of two pools,

Echo Rock Hot Springs ROBERT SHAVER

with an upper pool hovering around 106 degrees F and a lower pool slightly cooler at 103 degrees F. The pools are about 10 feet across and can easily hold several soakers. The pools may be a bit shallow for total submersion, but if you stretch out, you can get most of your body underwater. The upper pool has been constructed with rocks and mortar to create a nice soaking spot. The upper pool may be too hot for soaking, so you may want to divert the spout that fills the pool to let the water cool a bit before soaking.

37. GREELY BAR HOT SPRINGS

See map on page 173.

General description: A popular soak for the rafting crowd on the banks of the Owyhee River.

Location: Southeast Oregon, along the Owyhee River, about 45 miles downstream from the BLM Rome Boat Launch.

Development: None.

Best time to visit: Apr through June is probably the best time to visit Greely Bar Hot Springs, since water levels in the Owyhee River may drop too low after that to allow you to reach the hot springs by raft. You'll likely be sharing the soaking pool with other rafters during this time.

Restrictions: Camping is restricted within 200 feet of the hot springs.

Access: If you're rafting down the Owyhee River, then Greely Bar Hot Springs is an easy 1-mile walk from one of the 2 campgrounds above and below the hot springs. It's really not practical to visit Greely Bar Hot Springs overland, unless you're equipped for all-terrain travel and bushwhacking.

Water temperature: The soaking pool is almost too hot for comfort, often exceeding 110 degrees F. During the high water runoff in the spring, the Owyhee River may rise enough to mix with some of the hot springs water in the soaking pool, which may make for a more comfortable soak.

Nearby attractions: Geological and natural wonders are the hallmark of any trip down the Owyhee River.

Wildlife is abundant, including beaver, raptors, bighorn sheep, and even a herd of wild horses above the canyon rim. Lava flows and lava tubes are popular hikes, and petroglyphs are commonly found along the Owyhee River canyon.

Services: None. The nearest services are in Jordan Valley, including restaurants, motels, and gas.

Camping: Two campgrounds are located near Greely Bar Hot Springs— one is about a half mile upstream, and one is a half mile downstream from the springs. Camping is limited to 1 day and 1 night per river trip to help reduce pressure on the fragile desert ecosystem. Camping is prohibited within 200 feet of the hot springs. The Greely Bar campsites are one of the most popular (and most scenic) camping areas along the Owyhee River, so you'll probably be sharing the campground and hot springs with other rafting parties.

Map: BLM Owyhee Canyon Country Vale District South map.

GPS coordinates: N43.2095' / W117.5434'

Contact info: Bureau of Land Management, Vale District Office, 100 Oregon St., Vale, OR 97918; (541) 473-3144; www.blm.gov/office/vale-district-office.

Finding the springs: Rafting the Owyhee River is the best way to find Greely Bar Hot Springs. The hot springs and nearby campgrounds are about 45 miles downstream from the BLM Rome boat launch.

Greely Bar Hot Springs DEBORAH CASTORINA

THE HOT SPRINGS

Greely Bar Hot Springs is another of the pristine soaks along the Owyhee River. Greely Bar is similar to Ryegrass Hot Springs in that both are very near the riverbank, and both are often a little too hot for soaking unless cooler river water is added to the soaking pool. The campsites near Greely Bar are very popular during rafting season, so be prepared to share the soaking pool with other rafters.

38. RYEGRASS HOT SPRINGS

See map on page 173.

General description: A pair of hot but shallow soaking pools on the banks of the Owyhee River, popular with rafters when the river level is high in the spring and early summer.

Location: Southeast Oregon, in the Owyhee River canyon between the Rome and Leslie Gulch boat ramps.

Development: Some minor improvements have been made to the soaking pool by volunteers, but otherwise the area is pristine.

Best time to visit: Apr through June is the best time to visit Ryegrass Hot Springs, since water levels in the Owyhee River may drop too low after that to allow you to reach the hot springs by raft. You'll likely be sharing the pools with other rafters during this time.

Restrictions: Camping is prohibited within 200 feet of the hot springs.

Access: If you're rafting down the Owyhee River, then Ryegrass Hot Springs is an easy walk from the Ryegrass Campground. Reaching Ryegrass Hot Springs overland is not recommended—it's a difficult drive and scramble.

Water temperature: The hot springs emerge from the ground at 110 degrees F. The 2 shallow soaking pools are about a hundred yards downhill from the springs. The hot water in the pools averages around 105 degrees F. You may need to add some cooler river water to the pools to cool them down to an acceptable soaking temperature (so consider bringing a bucket with you).

Nearby attractions: Geological and natural wonders are the hallmark of any trip down the Owyhee River. Wildlife is abundant, including beaver, raptors, bighorn sheep, and even a herd of wild horses above the canyon rim. Lava flows and lava tubes are popular hikes, and petroglyphs are commonly found along the Owyhee River canyon.

Services: None. The nearest services are in Jordan Valley, including restaurants, motels, and gas.

Camping: A small campground is located near the riverbank close to Ryegrass Hot Springs. Camping is limited to 1 day and 1 night per river trip to help reduce pressure on the fragile desert ecosystem. Camping is prohibited within 200 feet of the hot springs.

Map: BLM Owyhee Canyon Country Vale District South map.

GPS coordinates: N43.0716' / W117.6973'

Contact info: Bureau of Land Management, Vale District Office, 100 Oregon St., Vale, OR 97918; (541) 473-3144; www.blm.gov/office/vale-district-office.

Finding the springs: Rafting the Owyhee River is the best way to find Ryegrass Hot Springs. The campgrounds near the springs are about 24 miles downstream from the BLM Rome boat launch.

THE HOT SPRINGS

The two small soaking pools at Ryegrass Hot Springs may not be as impressive in size as the other hot springs along the Owyhee River, but they still provide a welcome soak

Ryegrass Hot Springs DEBORAH CASTORINA

for weary rafters after a day of paddling. The soaking pools are close to the riverbank and may be a bit too hot for a comfortable soak without adding some cooler river water. The rock-lined pools provide a gorgeous view of the Owyhee River valley and surrounding canyons.

39. THREE FORKS WARM SPRINGS

See map on page 173.

General description: A series of warm soaking pools and a waterfall near the confluence of 3 tributaries of the Owyhee River.

Location: Southeast Oregon.

Development: None. Although the warm springs are on private land, they are also part of the Wild and Scenic River corridor, so absolutely no development of any kind is allowed.

Best time to visit: River rafters often stop for a soak at Three Forks Warm Springs during the high water rafting season in the spring and early summer. Hikers accessing the warm springs from across the river should wait until the water level has dropped in the late summer or fall (and even then the water current may be too strong to safely ford the Owyhee River). The 95-degree-F warm springs are probably not warm enough to justify a winter visit.

Restrictions: Three Forks Warm Springs is on private property, but the landowner has had a good relationship with rafters and hikers who stop by for a soak, so no restrictions have been imposed. Nudity is common in the soaking pools. High-clearance vehicles are recommended for the drive from Jordan Valley to the boat launch area. Keep your eye on the weather forecast—roads can be impassable after a rainstorm.

Access: The warm springs are located on private land, but the current landowner has imposed no restrictions on soaking. The surrounding land is managed by the BLM.

Water temperature: The water stays a consistent 95 degrees F. This is a bit cool for a winter soak but feels refreshing on a hot day in the summer or fall.

Nearby attractions: Geological and natural wonders are the hallmark of any trip down the Owyhee River. Wildlife is abundant, including beaver, raptors, bighorn sheep, and even a herd of wild horses above the canyon rim. Lava flows and lava tubes are popular hikes, and petroglyphs are commonly found along the Owyhee River canyon.

Services: The nearest services are in Jordan Valley, including restaurants, motels, and gas.

Camping: Four campsites are available at the BLM Three Forks boat launch area.

Maps: BLM Owyhee Canyon Country Vale District South map; USGS Three Forks, OR.

GPS coordinates: N42.5302' / W117.1839'

Contact info: Bureau of Land Management, Vale District Office, 100 Oregon St., Vale, OR 97918; (541) 473-3144; www.blm.gov/office/vale-district-office.

Finding the springs: From Jordan Valley head west along US 95 for 16 miles to a sign for Three Forks and the Soldier Creek Watchable Wildlife Loop. Turn left at the Three Forks sign and drive on a dirt road for another 35 miles south, following signs on Lower Soldier Creek Road until you reach the rim of the canyon overlooking the Owyhee River. From there it's a steep 2-mile descent to the river and the BLM Three Forks boat launch area. To reach the warm springs from the boat launch, you'll need to ford the Middle

Owyhee River. Only attempt to cross the river on foot during low water (otherwise take a raft or boat across). Once you are safely across, walk alongside the river up the canyon on a faint path for about 2 miles. Look up to your right and you'll see a waterfall that drops into the soaking pools. Scramble up to the pools and reward your efforts with a long soak.

THE HOT SPRINGS

Three Forks Warm Springs is one of the most scenic thermal spots in Oregon. The gorgeous views of the Owyhee River canyon coupled with the deep soaking pools and waterfall truly make this worth the effort. Over a thousand gallons per minute of 95-degree-F water forms a warm-water creek and waterfall. The largest pool is just above the waterfall—it can easily hold six or more soakers and is over 5 feet deep.

Three Forks Warm Springs KARL HESLER

APPENDIX: HONORABLE MENTIONS

The hot springs listed in the previous sections all provide wonderful soaking opportunities. But there are also several other hot springs that for various reasons aren't open for public bathing. In some cases hot springs pools and resorts that were once open to all have been purchased by owners who closed the bathing options to the public (and in some cases have even removed the soaking pools entirely). And some old spas have fallen into ruin, their empty pools and crumbling walls sitting quietly in open fields that were once filled with the laughter of soakers in decades past.

The four additional hot springs listed below no longer provide a soothing soak to visitors, but they have such interesting histories that it's worth including them in this guide (after all, this is a *touring* guide to hot springs, and not just a *soaking* guide!). If you happen to be near one of these closed hot springs, feel free to stop on public roads that provide a view of the closed resorts or old ruins. Savor the history from a distance, but don't trespass on private property for a closer look.

A. CRUMP GEYSER

General description: A slumbering hot-water well that was at one time the site of the largest continuously erupting geyser in the United States.

Location: South-central Oregon, 3 miles north of Adel.

Development: Three geothermal exploration wells were drilled across the road from the Crump Geyser well in 2012. These wells have intersected 265-degree-F thermal waters. Exploration has continued over the past decade, with plans to develop a geothermal power plant if sufficient water flow and temperature are found.

Best time to visit: Although the well that produces Crump Geyser has a fascinating history, it's probably not worth a special trip just to see an enclosed wellhead. Nevertheless, if you're visiting Hart Mountain National Antelope Refuge or Hunter's Hot Springs in Lakeview, consider looping south through the Warner Valley to take a look at the well. The road past Crump Geyser is open year-round, so you can visit any time of the year.

Restrictions: No Trespassing signs are posted on the fence enclosing the well at Crump Geyser. You can park your car about 50 yards from the well, but that's about as close as you can approach.

Access: Any vehicle can drive on paved CR 3-10 to see the hot-water well.

Water temperature: The water temperature in the well has been measured at 220 to 230 degrees F.

Nearby attractions: CR 3-10 (Hogback Road) continues north past Crump Geyser through the Warner Valley to the town of Plush. The Hart Mountain National Antelope Refuge (which contains Antelope Hot Springs) is about 20 miles northeast of Plush.

Services: None available at the site of the Crump well. Drive 3.3 miles south to Adel for food, gas, and water at the Adel Restaurant. All other services are available in Lakeview.

Camping: Drive north on CR 3-10 through Warner Valley to the Hart Mountain Antelope Refuge for overnight camping and RV spots (as well as a nice soak in Antelope Hot Springs). Other accommodations are available in Lakeview.

Maps: Oregon State Highway Map; *DeLorme: Oregon Atlas & Gazetteer,* page 85, E6.

GPS coordinates: N42.2265' / W119.8819'

Contact info: Ormat Technologies Inc., 6140 Plumas St., Reno, NV 89519-6075; (775) 356-9029; www.ormat.com.

Finding the springs: From Lakeview drive north for 4 miles on US 395 to the junction with OR 140. Turn east onto OR 140 and drive about 30 miles to the small community of Adel. Turn north onto CR 3-10 (also called Hogback Road) and drive 3.3 miles. On the west side of the road, look for a No Trespassing sign on a fence. Park your car near the sign. About 50 yards west of the sign, you'll see a mound of earth next to an 8-by-8 foot enclosure. Inside the enclosure is the wellhead for Crump Geyser.

About 200 yards east of the well (across the county road) are some small remnants of long-extinct geyser cones and other mineral deposits that indicate geothermal activity in past years. You'll also see signs indicating the geothermal exploration activities underway since 2012.

History

In the early 1950s a rancher named Charles Crump owned 1,300 acres of hay and pastureland in the barren Warner Valley east of Lakeview. In 1959 Crump decided that he needed a new source of irrigation water, so he drilled a well on the west side of the valley. To his surprise the well hit a pocket of hot steam and began erupting every few hours.

Crump was intrigued with the commercial potential of his steaming well; he knew that similar wells had been drilled in California and were being used to power electrical generation plants. Four years after drilling his initial well, Crump contracted with Magma Power Company of Los Angeles to drill a larger and deeper well near his little geyser, hoping to find 300-degree-F steam that could power a steam turbine for an electrical generator.

In the summer of 1959, Magma Power arrived at Crump's ranch and drilled a 20-inch-diameter well more than 1,600 feet deep. The drillers recorded water and steam temperatures of 250 degrees F in the well, but the 1950s technology for converting geothermal steam to electricity required even hotter temperatures (near 300 degrees F). Disappointed with the water temperature, the energy company removed the drill rig and headed back to California.

Thinking the well was a failure, Crump returned to his ranching. Two days after the well had been abandoned, though, it roared to life, erupting a plume of steam and water that rose more than 150 feet into the air at a continuous volume of 400 and 500 gallons per minute. Unlike the little geyser that Crump had drilled a few years earlier, this new gusher of steam and hot water erupted continuously for more than nine full months.

It must have been an amazing experience to drive through the barren Warner Valley and see the great geyser roaring out of Crump's well. The *Oregonian* sent a reporter in January 1960 to report on this geologic wonder:

Nature has her wires crossed in a quiet valley some 30 miles northeast of Lakeview, and the result is a spectacular nightmare of ice and steam that few folks are likely ever to see.

The scene is a small patch of land in a corner of the Warner Valley on the ranch of Charles A. Crump. This is Crump's Geyser, a gusher of water, boiling from the earth at 220 degrees in a stream that spouts 150 feet into the air, at least 30 feet higher than Old Faithful.

The icy wind blowing down from the Warner Rim has caught up the steam, freezing some of it before it hits the ground and creating a three-acre sea of ice that ripples over the field and roadway in one great slippery blanket. Taut fence wire has been crushed to the ground by the heavy sheets of ice built up by the freezing vapor from the geyser, fence posts have been transformed into marble-like columns, almost alive with the layer upon layer of ice built up on the freezing weather.

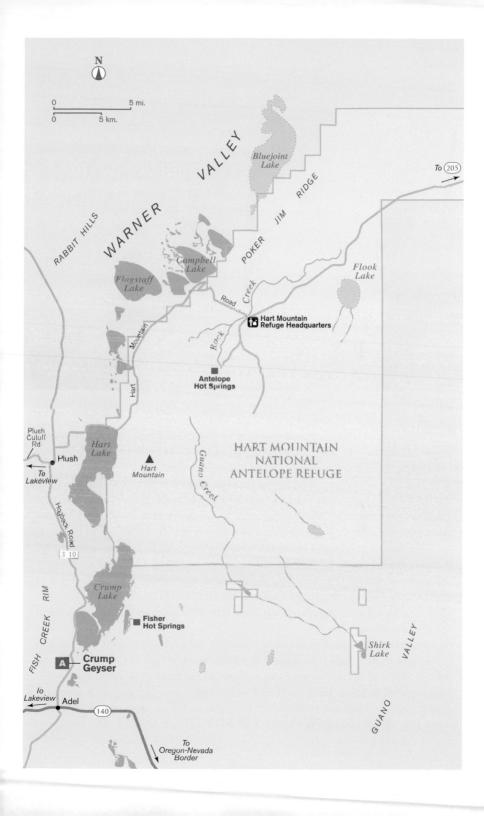

N

0 ____ 5 mi.
0 ____ 5 km.

WARNER VALLEY

RABBIT HILLS

Bluejoint Lake

POKER JIM RIDGE

To 205

Campbell Lake

Flagstaff Lake

Flook Lake

Road

Rock Creek

Hart Mountain Refuge Headquarters

Mountain

Hart

Antelope Hot Springs

HART MOUNTAIN NATIONAL ANTELOPE REFUGE

Plush Cutoff Rd

Hart Lake

Plush

Hart Mountain

Guano Creek

To Lakeview

Hogback Road

3 10

Crump Lake

FISH CREEK RIM

Fisher Hot Springs

A Crump Geyser

Shirk Lake

GUANO VALLEY

To Lakeview

Adel

140

To Oregon-Nevada Border

Geothermal exploration near Crump Geyser

The geyser may have continued to spout for a very long time, but vandals clogged the well with boulders and rocks in the spring of 1960. The wondrous Crump Geyser has never erupted since.

The story doesn't end there, however. The original irrigation well that Crump had drilled five years earlier ceased erupting when the massive 1959 geyser sprang to life. After the vandalism halted the eruption of the new geyser, Crump's "old" geyser began to spout again intermittently, erupting 90 feet into the air every 8 to 10 hours.

Through the 1960s and 1970s, the "old" geyser continued its infrequent display, but over time the intervals increased between eruptions, and eventually the geyser ceased altogether. As late as the mid-1990s, the Crumps could occasionally force an eruption by lowering a bucket into the well until it filled with water and then retrieving it, an action that disturbed the dynamics of the water pressure just enough to cause the geyser to blow.

Interest in the Crump well and surrounding areas revived around 2010, when a geothermal energy company invested money in new exploration wells searching for hotter water and steam. At least three exploration wells have been drilled across the road from the original Crump well, and temperatures over 260 degrees F have been found. Further research and exploration are planned, with expectations of finding water temperatures exceeding 300 degrees F that could be used in a geothermal power plant to produce electricity.

B. BLUE MOUNTAIN HOT SPRINGS

General description: A peaceful, warm-water swimming pool on the grounds of a former guest ranch. The pool has been closed to the public since the early 2000s.

Location: Northeast Oregon, 10 miles southeast of Prairie City near the Strawberry Mountain Wilderness.

Development: The hot springs have been home to a hotel and swimming pool for well over a century, although the public is not currently allowed.

Best time to visit: The hot springs swimming pool is currently (as of 2020) closed to the public, so there's no good time to visit.

Restrictions: None; closed to the public.

Access: Any vehicle can make the 10-mile trip along the blacktop road from Prairie City to the road to the hot springs. But it's likely you'll encounter a No Trespassing sign before you leave the main road.

Water temperature: The hot springs average 136 degrees F, cooling to 120 degrees F as the water enters the swimming pool 10 feet to the north. The temperature of the swimming pool hovers around 100 degrees F.

Nearby attractions: The 1.5 million-acre Malheur National Forest surrounding Blue Mountain Hot Springs is home to many recreational opportunities. The forest offers camping, fishing, snowmobiling, hunting, and other outdoor opportunities. Strawberry Mountain Wilderness attracts backpackers who explore the alpine lakes surrounding the peak of 9,038-foot Strawberry Mountain. Malheur National Forest was in the world spotlight in the late 1990s when scientists announced that they had discovered the largest organism on earth in the forest near Prairie City. The organism, a subterranean mushroom named *Armillaria ostoyae*, covers an area more than 3 miles across (that's more than 1,500 football fields) and extends an average of 3 feet into the ground. This monster mushroom has probably been growing for close to 2,500 years, slowly extending its filaments through tree roots beneath the forest floor. In autumn small brown mushrooms pop up to the surface after rain showers, but the threadlike filaments that comprise the bulk of the organism remain hidden beneath the surface. Scientists from Oregon State University discovered the mushroom after looking at aerial photos of large expanses of trees killed by root rot caused by the mushroom. Stop by the Forest Service ranger station in Prairie City for more information on this botanical behemoth.

Services: The Prairie City area offers some unique accommodations, including the Riverside Schoolhouse Bed & Breakfast (541-820-4731, www.riversideschoolhouse.com), which is 4 miles north of the hot springs on CR 62, and the Hotel Prairie (541-820-4800, www.hotelprairie.com), located within the city limits of Prairie City.

Camping: Several campgrounds are nearby, including Depot Park (which is the city park in Prairie City). Depot Park has facilities for both tent camping and RVs. There are also several campgrounds south of Blue Mountain Hot Springs in the Malheur National Forest and Strawberry Mountain Wilderness. Stop at the Forest Service office in Prairie City for current camping information and regulations.

Maps: Oregon State Highway Map; *DeLorme: Oregon Atlas & Gazetteer,* page 78, D3.

GPS coordinates: N44.3548' / W118.5748'

Contact info: Blue Mountain Hot Springs, Star Route, Prairie City, OR 97876; (541) 820-3744.

Finding the springs: From US 26 in Prairie City, turn south onto Main Street. Drive 0.3 mile, then turn right onto Bridge Street. Continue south on Bridge Street across the John Day River. Bridge Street then turns into CR 62 (also called Logan Valley Road or South Side of River Road). Drive 10 miles southeast on CR 62. Between mileposts 10 and 11, look for a large black mailbox and a private property sign on the north side of the road. If the hot springs are still closed to the public, you may have to turn around at this point without going up about one-third mile past an old homestead to the 2-story Blue Mountain Lodge.

THE HOT SPRINGS

The open-air concrete swimming pool has changed little in the past half century. The pool is about 4 feet deep and measures 30 feet by 70 feet. The actual hot springs bubble into a small basin 10 feet west of the swimming pool on the opposite side of a chain-link fence. Plastic pipes bring the 120-degree-F natural hot water into the pool. The pool temperature averages about 100 degrees F. The warmest soaking spot is near the concrete steps close to the pipe where the main hot springs enter the pool.

History

The hot springs were first claimed in the 1860s by homesteader John Douglas, who built a log house and an enclosed swimming pool. Blue Mountain Hot Springs became a popular vacation destination, and subsequent owners in the early 1900s built a hotel, a dance hall, a horse stable, a bathhouse, and an icehouse.

E. C. Tuttle purchased the hotel and hot springs in 1945. Tuttle, who made his fortune as an executive with the Winchester Firearms and Ammunition Company, closed the springs to public access. From 1945 until 1965 Tuttle used the property as a private retreat for his family and friends.

Testimonies abound as to the healing properties of the mineral water at Blue Mountain Hot Springs. In 1945 a construction worker spent five months at the springs, performing carpentry work on Blue Mountain Lodge for the new owner. More than two decades after his summer at the springs, the construction worker wrote a letter to the current owners of the resort, describing the wonderful benefits he received from the hot springs water:

My Dear Sir:

I am enclosing a picture of the Blue Mountain Lodge as it looked when I arrived there April 18, 1945 and also a picture of the completed work taken on September 20, 1945. Everybody was pleased with outcome including the

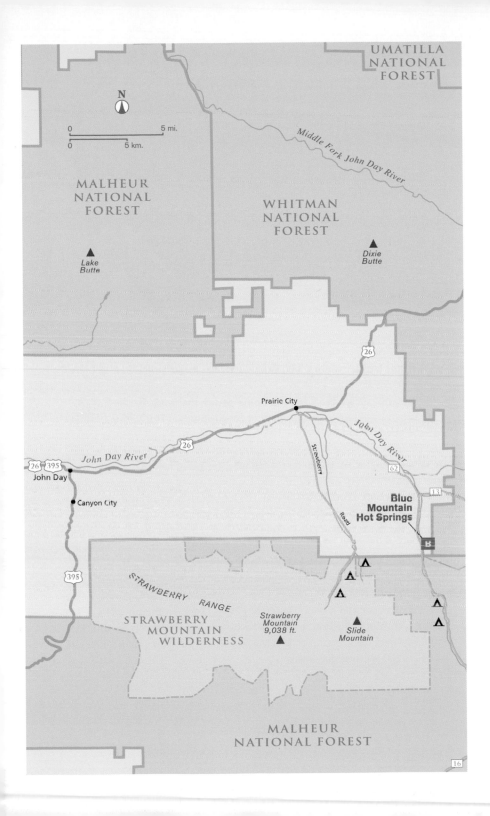

N

0   5 mi.

0   5 km.

UMATILLA
NATIONAL
FOREST

Middle Fork John Day River

MALHEUR
NATIONAL
FOREST

WHITMAN
NATIONAL
FOREST

▲
*Lake
Butte*

▲
*Dixie
Butte*

26

Prairie City ●

John Day River

26

26 395

John Day

● Canyon City

Strawberry

John Day River

62

13

**Blue
Mountain
Hot Springs**

Road

395

STRAWBERRY RANGE

STRAWBERRY
MOUNTAIN
WILDERNESS

*Strawberry
Mountain
9,038 ft.*
▲

▲
*Slide
Mountain*

MALHEUR
NATIONAL FOREST

16

The swimming pool at Blue Mountain Hot Springs

owner Mr. Tuttle, my boss, Mr. Stanton, the architect and myself as supervising the architects.

You know I received a reward for my part in this work, besides my pay.

When I went up to do this work at Blue Mountain Hot Springs, I didn't feel too good. I had been troubled with arthritis in my hands, feet, legs, arms and shoulder for some time—about eight years or more. I normally weighed 165 pounds and when I went on the job up there I weighed 142 pounds. Anyway, after about six weeks on the job I noticed I was feeling better and my joints didn't hurt as much. I had been drinking the hot water from the spring from then until fall or September. I sure filled up on the hot water and cold water from the irrigation ditch from Rail Creek. I expect I drank about a gallon or more of hot water each day. That fall and before, the swelling in my hand and my hurting joints had all gone away. I helped the caretaker of the ranch split his wood after living there a month. I went home without an ache or pain and I weighed 172 pounds. I have never had any arthritis since—that was my extra pay and bonus. I am grateful that I took on the supervisor's job at the Blue Mountain Hot Springs and I hope some other person with that trouble would try it out. I think it would work even if they didn't have the same kind of arthritis that I had.

Yours truly,

Charles R. Kaufman
Portland, Oregon

Upon Tuttle's death in 1965, Eugene and Helen Ricco purchased Blue Mountain Hot Springs and reopened the swimming pool to the public. The Riccos also reopened the Blue Mountain Lodge and provided room and board to guests. The Ricco family still owns the pool and adjacent ranch, although the hotel and pool are no longer open to the public.

C. LEHMAN HOT SPRINGS

General description: A popular mountain resort that featured one of the largest pools in the Pacific Northwest. The resort has been closed to the public since 2009.

Location: Northeast Oregon, 38 miles west of La Grande in the Blue Mountains.

Development: First developed in the 1880s as a summer resort, Lehman Hot Springs was an all-season destination for outdoor enthusiasts.

Best time to visit: The resort was busy year-round until it closed to the public in 2009. Families often rented the large bunkhouse for reunions. Mountain biking was popular in summer, hunters used the resort as a base in fall, and winter brought a migration of snowmobilers and cross-country skiers. Hopefully these activities will once again become available if the resort reopens.

Restrictions: Bathing suits were required in the pools when the resort was open.

Access: Any vehicle can make the trip on the paved highway and gravel road that lead to the hot springs. A No Trespassing sign and barrier is in place just before reaching the resort, so visitors can't enjoy the resort unless it reopens to the public.

Water temperature: The hot springs average a steamy 168 degrees F at their source on the hillside above the pools. The thermal soaking pool temperatures vary with the season, ranging from 92 to 130 degrees F.

Nearby attractions: The hot springs are a private inholding surrounded by the Wallowa-Whitman National Forest. East of the resort, Hilgard State Park offers camping and picnic spots along the banks of the Grande Ronde River. Close by is the Anthony Lakes Mountain Resort, featuring both downhill and cross-country skiing. West of the hot springs, the Forest Service maintains the Winoma-Frazier Off-Highway Vehicle Complex, which provides motorized-vehicle enthusiasts with more than 100 miles of trails.

Services: If the hot springs ever reopen to the public, a variety of on-site lodging may be available. A deli and gift shop were located next to the swimming pools, but these are currently closed to the public, as is the entire resort.

Camping: Camping is available at the Forest Service's Frazier Campground, situated 2 miles east of Lehman Hot Springs. To reach the campground, drive 2 miles east on OR 244, then turn south onto FR 5226 for 0.5 mile.

Maps: Oregon State Highway Map; *DeLorme: Oregon Atlas & Gazetteer,* page 74, F2.

GPS coordinates: N45.1509' / W118.6601'

Contact info: Lehman Hot Springs, Box 187, Ukiah, OR 97880; (541) 427-3015; www.lehmanhotsprings.org.

Finding the springs: From La Grande drive west on I-84 for 7.5 miles. Turn south onto OR 244. Drive southwest for 30 miles to milepost 17. Turn south onto Lehman Springs Road (CR 917) for 1.5 miles to the hot springs resort. If you're heading from Ukiah, drive 16 miles east on OR 244 to milepost 17 and turn south. You'll likely be stopped by a closed gate posted with a No Trespassing sign. Hopefully the resort will be reopened to the public in the future.

THE HOT SPRINGS

The steaming valley first discovered by James Lehman contains more than fifty separate hot springs that bubble to the surface 500 yards uphill from the swimming pools. The runoff from these springs forms a hot-water creek that flows parallel to a cold-water creek a few feet away. Legend has it that visitors to the hot springs would fish for trout in the cold stream. Upon catching a dinner-sized fish, the angler would swing the hooked fish into the adjacent hot-water creek, where it would cook in a few minutes.

The largest of the many hot springs is collected into pipes that lead downhill to the swimming-pool complex. Hot water first enters the smallest and hottest pool, which reaches a maximum temperature of 130 degrees F in summer, cooling to around 110 degrees F in winter. First-time bathers sometimes try to soak in this pool, but it's almost always too hot. The middle pool is a more comfortable 106 to 108 degrees F, whereas the lower pool stays between 92 and 96 degrees F. The temperature in the cold-water pool hovers around 55 degrees F. The four pools combined exceed 9,000

No Trespassing signs on the road to Lehman Hot Springs

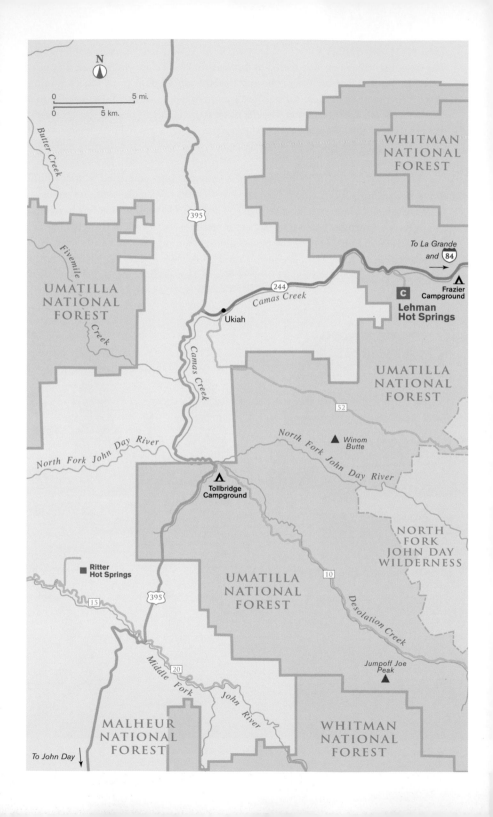

N

0   5 mi.

0   5 km.

Butter Creek

WHITMAN
NATIONAL
FOREST

395

Fivemile Creek

UMATILLA
NATIONAL
FOREST

To La Grande
and 84

244

Camas Creek

Ukiah

C

Frazier
Campground

Lehman
Hot Springs

Camas Creek

UMATILLA
NATIONAL
FOREST

52

North Fork John Day River

Winom
Butte

North Fork John Day River

Tollbridge
Campground

NORTH
FORK
JOHN DAY
WILDERNESS

Ritter
Hot Springs

UMATILLA
NATIONAL
FOREST

10

Desolation Creek

15

395

Jumpoff Joe
Peak

Middle Fork

20

John River

MALHEUR
NATIONAL
FOREST

WHITMAN
NATIONAL
FOREST

To John Day

square feet, which resort literature claims is the largest swimming pool complex in the Pacific Northwest.

History

Legend has it that the Umatilla and Cayuse tribes summered in the area near the hot springs to pick huckleberries. James Lehman and Dr. John Teel were the first Europeans to visit the hot springs. While hunting in the area in the fall of 1871, Lehman and Teel followed some deer tracks into a small canyon, where they discovered the steaming hot water. The men camped near the springs overnight, then traveled to Pendleton the next day to file a claim on the land. The hot springs were called "Teel Springs" until the death of Dr. Teel in 1880, when James Lehman became the sole owner and namesake.

Within a few years Lehman turned the natural springs into a popular summer resort, complete with hotel, store, barbershop, dance hall, and swimming pool. According to a history of the resort written by Mildred Searcey, hot mud baths were available in small cabins built next to the pool, where guests would soak in "slimy, gooey mud."

The twenty-eight-room hotel that Lehman built was eventually destroyed by fire, and the other original resort buildings were replaced. Patrick and Rachel Lucas purchased the resort in the 1990s. Water quality became an issue in 2009, and the resort was closed to the public by court order. Fancho "Fee" Stubblefield purchased the resort in 2012 and has been working to bring the resort back up to code and reopen it. (Fancho's grandfather, also named Fancho, had purchased Lehman Hot Springs in 1925 for $500, and the Stubblefield family has had a long association with the property.)

D. MEDICAL SPRINGS

General description: A pioneer resort that featured an Olympic-size swimming pool and large hotel. The pool and hotel have been closed to the public for decades.

Location: Northeast Oregon, 35 miles southeast of La Grande in the foothills of the Wallowa Mountains.

Development: The hot springs have been used commercially since the first bathhouse was built in 1869. The hot water is currently used for heating a private lodge and an Olympic-size swimming pool. Unfortunately the pool is off-limits to the public, but it's worth stopping for a quick look.

Best time to visit: OR 203 passes right by Medical Springs, so you can drive through the area any time of the year. You can observe the large swimming pool and old hotel from the road.

Restrictions: The pool, hotel, and general store are closed to the public.

Access: You can easily view the pool, hotel, and old general store, since OR 203 runs within a few yards of all these structures. But stay on the roadside when observing.

Water temperature: The hot springs emerge from the ground at 140 degrees F and then are piped 200 yards to the 50-by-150-foot swimming pool, where the water is cooled to around 104 degrees F.

Nearby attractions: The charming Victorian-era town of Union is 20 miles northwest of Medical Springs. Union County Museum has excellent cowboy and geology exhibits, as well as period rooms. The Eagle Cap Wilderness northeast of Medical Springs offers several hiking trails to high mountain lakes. Consider driving the 95-mile-long, figure-eight Grande Tour from La Grande to Hot Lake Springs and on to Union and Medical Springs. The tour then circles back to Union and north to Cove before returning to La Grande. This is a good daylong tour if you're staying in La Grande or Union. In addition to Medical Springs, you'll see the hot springs and hotel at Hot Lake Springs, the soaking pool at Eagles Hot Lake RV Resort, and the warm-water swimming pool in Cove. Pick up a tour map and area highlight descriptions at the La Grande/Union County Visitors and Convention Bureau.

Services: None available at the springs. A small general store at one time operated near the swimming pool, but this has been closed for several years. You'll need to drive to Union or La Grande for gas, groceries, and restaurants. For a historical overnight lodging experience, stay at the Union Hotel in the nearby town of Union. The hotel was built in 1921 in the American Renaissance style. Call (541) 562-1200 or visit www.theunion hotel.com.

Camping: There's a nice creek-side campground in Catherine Creek State Park, 12 miles northwest of Medical Springs on OR 203.

Maps: Oregon State Highway Map; *DeLorme: Oregon Atlas & Gazetteer,* page 75, F7.

GPS coordinates: N45.0166' / W117.6292'

Finding the springs: From La Grande drive 15 miles south on OR 203 to the town of Union. Continue through Union southeast on OR 203 for an additional 20 miles to Medical Springs. Park your car on the side of the road across from the swimming pool and take a look over the fence at the historic bathing spot.

History

Medical Springs is inextricably linked to Dunham Wright, one of the most influential frontier politicians in eastern Oregon. Originally from the Midwest, Wright moved to Cove, Oregon, in the early 1860s. Barely 20 years old, Wright was anxious to purchase his own land. A rancher in the area told Wright about some unclaimed property southeast of Cove that had hot springs, and Wright rode out to investigate. Years later, in his memoirs, Wright described his first view of Medical Springs and the Native Americans camped nearby:

> The springs were located in a big willow grove. The men would build a number of small dams across the streams that ran from the springs. The water

The old swimming pool at Medical Springs

Road sign outside Medical Springs

would accumulate to the depth of about twenty inches. Sticks were placed around the edge, and then a big elk hide or blanket would be stretched across the top to keep the steam in. The men would then crawl in like "dogs in a kennel." When they were done steaming, they would jump into the cold water creek, which they had dammed to about three feet deep. This was thirty yards from the hot springs (where the hotel later stood). The final stage was to get out of the water, wrap themselves in warm blankets, then "lay down to almost melt in their tepees."

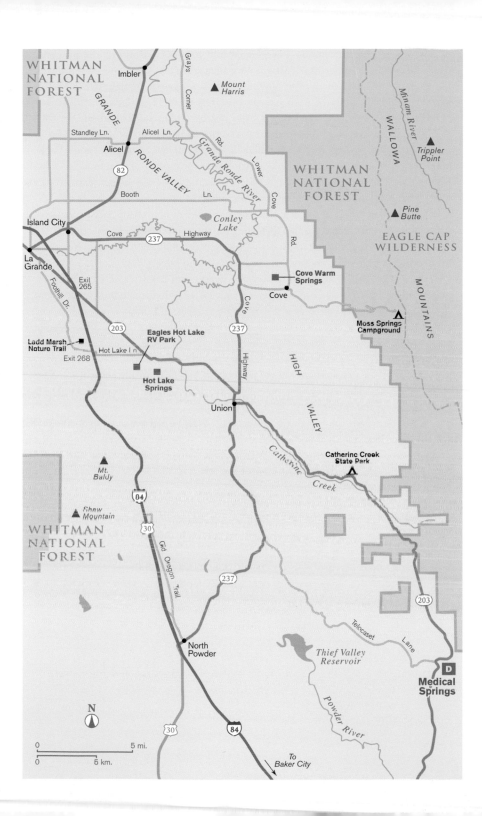

Wright's bride, Artemesia, encouraged him to stake a claim to the property, which he did in December 1868. Wright built a log cabin and a small log bathhouse on the property in 1869. Shortly after moving to Medical Springs, Wright became active in state politics and served for ten years in the Oregon State Legislature.

The little resort attracted visitors from nearby Union and La Grande. In 1886 Wright built his first hotel. The two-story structure measured 26 by 100 feet and contained forty guest rooms. In 1905 he added a sanitarium and additional rooms to the hotel, as well as a livery stable, a general store, and a post office. The remodeled hotel featured two parlors, "one with a pool table for the men, and the other a piano for the women." A large maple-floored ballroom occupied the upper floor.

Wright was proud of the beautiful surroundings of his resort. A brochure produced to market the resort claimed that "the pure mountain air, the grand scenery, the excellent fishing and hunting are valuable adjuncts in tempting the invalid out of his morbid self, and fostering in him renewed courage and hope."

Unlike the fancy hotel and sanitarium at nearby Hot Lake, which attracted well-heeled visitors, Wright's resort attracted a rough-edged crowd, including "miners with rheumatism, gamblers who played a hard game of cards and smoked old stogies, and rowdy, drunken cowboys over-zealous with their six-shooters." For a time Wright sponsored a rodeo at the resort. Apparently the popular event ended when, according to his grandson, Wright realized that he couldn't control the crowd: "too many drunken cowboys and only one sheriff per county." Even normally genteel activities at the resort were boisterous. One newspaper article noted that "irate croquet players were known to end a match by shooting at the balls."

A fire destroyed the wooden hotel in 1917. Wright had not adequately insured the structure and only had enough funds to build a smaller six-room hotel on the property the following year. In 1929 Wright built the open-air pool that is still used by his descendants. He continued to live at the resort until his death in 1942, shortly after his one-hundredth birthday.

Medical Springs closed to the public in the 1950s when a lumber mill in a nearby town went out of business. Most of the town residents who had frequented Medical Springs were thrown out of work and left the area, which caused a drastic reduction in resort revenues.

Wright's descendants still live in the old hotel at Medical Springs. The swimming pool is closed to the public; only Wright's family members and the occasional local scout troop swim in the waters.

BIBLIOGRAPHY

Belknap Resort & Hot Springs—An Historical Guide 1854–1996. McKenzie Bridge, OR: 1996.

Berry, George. "Thermal Springs List for the United States." NOAA Key to Geophysical Records Documentation No. 12. Boulder, CO: National Geophysical and Solar-Terrestrial Data Center, 1980.

Blinn, Mayme Schwartz, and Richard Joseph. *Upper John Day River—Early Days in Prairie City.* Prairie City, OR: privately printed, 1977.

Bloomquist, Elsie. "A History of Carson." *Skamania County Heritage* 14, no.1 (June, 1985).

Bloomquist, R. Gordon. *Geothermal Energy in Washington: Site Data Base and Development Status.* Klamath Falls, OR: OIT Geo-Heat Utilization Center, 1979.

Fiege, Bennye. *The Story of Soap Lake.* Soap Lake, WA: Soap Lake Chamber of Commerce, n.d.

Foster, Teresa. *Settlers in Summer Lake.* Bend, OR: Maverick Publications, 1989.

Hill, Lawrence D. "Tales from the Hills." Ontario, OR: *Daily Argus Observer*, 1982.

Horowitz, Howard. "The Landscapes of Hot Springs and Mineral Springs in Western Oregon." Master's thesis, University of Oregon, 1973.

Irving, Washington. *Astoria; or Anecdotes of an Enterprise Beyond the Rocky Mountains.* Philadelphia: Carey, Lea, & Blanchard, 1836.

Jackson, Royal G., and Jennifer A. Lee. *Harney County—An Historical Inventory.* Burns, OR: Harney County Historical Society, 1978.

Jensen, Veryl M. *Early Days on the Upper Willamette.* Oakridge, OR: Upper Willamette Pioneer Association, 1970.

Justus, Debra. *Geothermal Resources in Oregon: Site Data Base and Development Status.* Klamath Falls, OR: OIT Geo-Heat Utilization Center, 1979.

Korosec, M. A., et al. *Geothermal Resources of Washington—Geologic Map GM-25.* Olympia, WA: Washington Department of Natural Resources, 1981.

Peters, Shirley, and Sheila Smith. *Hot Lake—The Town Under One Roof.* La Grande, OR: privately printed, 1997.

Peterson, Norman V., et al. *Geothermal Resources of Oregon* [map]. Portland, OR: Oregon Department of Geology and Mineral Industries, 1982.

Searcey, Mildred. *Way Back When.* Pendleton, OR: East Oregonian Publishing Company, 1972.

Shaffer, Leslie L. D., and Richard Baxter. "Oregon Borax: Twenty Mule Team—Rose Valley History." *Oregon Historical Quarterly* 73, no. 3 (September 1972).

Simerville, Clara L., ed. *One Century of Life: Dunham Wright of Oregon 1842–1942.* Corvallis, OR: privately printed, 1986.

Southworth, Jo. "The Ritter Hot Springs." *Blue Mountain Eagle* (John Day, OR), 23 November 1972.

US Fish and Wildlife Service. *Recovery Plan for the Borax Lake Chub,* Gila boraxobius. Portland, OR: US Fish and Wildlife Service, 1987.

ONLINE SOURCES FOR
WASHINGTON AND OREGON HOT SPRINGS INFORMATION

Most hot springs listed in this book have their own websites (see individual chapters for web addresses). There are also some compilation websites that often have the latest access and updates on Oregon and Washington hot springs—check these for the latest chatter on individual hot springs before your next trip. Also check Facebook for interest groups around individual hot springs in Oregon and Washington as these often contain the most current information.

Facebook: Oregon's Hot Springs (A very active and up-to-date forum for the latest information on Oregon hot springs.)
www.facebook.com/oregonhotsprings

Hot Water Slaughter—Oregon Hot Springs
www.hotwaterslaughter.com/category/hotsprings/oregon

Hot Springs of Oregon—A Detailed Directory
http://oregonhotsprings.immunenet.com

Soakers Forum—A Place for Natural Hot Springs Resources
www.soakersforum.com

SoakOregon
SoakOregon.com

INDEX

ABOUT THE AUTHOR

Jeff Birkby's fascination with hot springs began in the early 1980s when he managed geothermal-energy projects for the Montana Department of Natural Resources. During his years of working for the state of Montana, Jeff developed a passion for hot springs lore, especially the stories and legends of hot springs in the northwestern United States. Jeff is a member of the Humanities Montana Speakers Bureau and often lectures on the social history of hot springs of the West. Jeff is also the author of the FalconGuide *Touring Hot Springs Montana and Wyoming*, as well as *Geothermal Energy in Montana—A Consumer's Guide*, published by the Montana Department of Environmental Quality. He also authored *Images of America—Montana's Hot Springs* through Arcadia Publishing in 2018.

Jeff consults with hot springs owners on how they can use their geothermal resources for greenhouse, pool, and space heating, as well as for electric power generation.

More than a dozen soakable hot springs are within a 2-hour drive of Jeff's home in Missoula, Montana.

If you have comments, corrections, or fresh information about any Washington or Oregon hot springs, send your insights to Jeff Birkby c/o Falcon Publishing, or e-mail Jeff at jeffbirkby@gmail.com. He will check out your information for future editions of this guide.